The Post-Pandemic World

John Erik Meyer

The Post-Pandemic World

Sustainable Living on a Wounded Planet

 Springer

John Erik Meyer
Canadians for a Sustainable Society
Parry Sound, ON, Canada

ISBN 978-3-030-91784-5   ISBN 978-3-030-91782-1   (eBook)
https://doi.org/10.1007/978-3-030-91782-1

This Springer imprint is published by the registered company Springer Nature Switzerland AG
The registered company address is: Gewerbestrasse 11, 6330 Cham, Switzerland

To the people who are able to consider future generations and who strive to pull their own ecological weight. Leaders make an effort before any is demanded of them.

Preface

Post-Pandemic World: Sustainable Living on a Wounded Planet offers a comprehensive strategy for investing your time and money to assure high levels of energy security and control over the cost of living. Reducing your carbon and ecological footprint builds your security and improves that of your children and community.

Human society has recently been overtaken by a calamity which, based on excess death estimates, has claimed more than ten million lives. The Covid-19 pandemic has brought the fragile nature of our globalized, just-in-time economic structure into the spotlight and exposed many of the vulnerabilities our society has to a simple biophysical shock.

During the pandemic, those who took an active interest in their health had better outcomes than those who did not. Those who lived in regions with better government response also did better. The Covid-19 experience should be a valuable learning experience for all of us.

The much larger threats of climate change and resource depletion present more complex challenges. Our efforts to mitigate and adapt to them will require a process involving far deeper changes taking place over many decades.

Transformation must be led by individuals. People need to have a sound overview to effectively invest their time and money to take control of their energy and carbon budgets. Securing their own resilience and capping their living costs will offer considerable protection against the probable energy, resource, and climate shocks which lie ahead.

Covid-19 served notice that we need to be more aware of our surroundings. It also allowed us to demonstrate that we are able to change. In reacting to the Covid-19 threat, we broke our fixation with the singular goal of growth in the size of the commercial economy. In this regard, the pandemic can be seen as a relatively cheap wake-up call to the existential threat of a planet under extreme stress.

Earth is beginning to exit the Goldilocks zone of nearly ideal climatic conditions in which humanity has blossomed over the past 12,000 years. The days of rich, seemingly endless cheap fossil fuel reserves are behind us. The path of climate stability and abundant energy and resources that allowed us to reach our current level of sophistication is not a path we will be able to follow in the future.

This book clearly presents these important concepts:

- How climate, energy, and resources have defined human history
- How to evaluate your housing, transport, and lifestyle options
- What you can do for yourself
- What you can do for your community and country
- The importance of walking into the future rather than being dragged into it
- The social awareness and the biophysical economic tools necessary to build a pathway to sustainability

Are we destined to act out a societal version of the Peter Principle whereby a society continues to grow in size and complexity until it fails? Or have we by now seen enough re-runs of this film to be able to avoid doing it one more time in a very big way?

The inevitable transition to renewable energy requires all individuals and organizations to adjust their consumption levels as well as the way they access and spend energy. This book is for individuals who want to take charge of this process for their own security and that of their families and communities. It will help people gain a perspective of how to invest their time and money to best assure a stable energy supply while minimizing long-term costs.

A full and healthy lifestyle does not need to be an environmentally destructive one. This is one lesson Covid-19 has taught us. We can apply that learning to transforming our relationship with the planet from one of desperate exploitation to one of nurturing it for a sustainable future.

Parry Sound, ON, Canada   John Erik Meyer

Acknowledgments

I'd like to thank all of the people who provided support through discussion, commentary, editing, and good wishes during the entire creation process. Certainly the patience and support of Ute Heuser and Amrita Unnikrishnan of Springer Nature were critical as were David Packer's conceptual insights. Gomathi Mohanarangan, thank you for the final fine tuning!

I thank Andrew Marshall and David Gascoigne who provided much needed editing and Sam Pratt and Joyce Sykes for their practical insights. Thomas Althouse and Peter Stubbins made clear suggestions on form and wording while Art Hunter served up mounds of information from his living lab. Forrest Pengra and Andre Aasrud contributed their insights into the practical application of new technology in the public policy arena.

Thank you once again to Charlie Hall who has remained a source of cogent advice and endless technical knowledge and to Chris Rhodes who was kind enough to re-acquaint me with the proper use of the English language.

Contents

About the Author

John Erik Meyer has been working with per capita resource issues for 5 decades and in the past 10 years has focused much of his work on the most important resource: energy. His first book, *The Renewable Energy Transition, Realities for Canada and the World* dealt with energy as a determining factor in the development of human society. He has worked with physical unit measurement, dollar accounting, and costing systems, and this has brought to light the limitations of the GDP measurement system his degree in economics taught him to focus on. Mr. Meyer is a patent holder, energy positive house builder (in progress), and small medium-tech business owner. He has had articles published in Canada's major newspapers dealing with population, immigration, and the failings of our GDP measurement systems, to reflect social and environmental reality. He created the www.perfectcurrency.org website to present a real value-based currency system as opposed to our current fiat and lately crypto currency systems. As well, he developed the sustainablesociety.com website to connect the dots between societal collapses of the past and the resource issues which triggered them. His goal is to see humanity develop the biophysical focus and tools necessary to establish an egalitarian and sustainability society on a healthy planet.

1

Society, Energy and the Natural World

Standing at the Top of the Hill

At this early point in the twenty-first century, humanity finds itself in complete dominance of the planet on which our ancestors struggled to survive for millennia. Once tied to the ebbs and flows of natural systems, our current activities are largely independent of the vagaries of the natural world due to the huge fossil fuel energy stores which have become available to us in the past 300 years.

These have enabled us to vastly expand our numbers, our level of material consumption and our exploitation of natural resources virtually without geographic restriction. We can go anywhere, live anywhere and do virtually anything we choose to with the energy now at our disposal.

Yet, even with our success and explosive accumulation of knowledge over the past 12 generations, the worldview we have now looks a great deal different through the front window than it does in the rearview mirror. On looking to the past, we see a small human population beginning to exploit a virgin planet. Now, despite our dominance over the natural systems which once constantly threatened our survival, the richness of the natural systems, upon which we still depend, is declining rapidly and the end of our once-in-a-species energy bonanza is in sight.

By the end of this century, we will need to have largely weaned ourselves off fossil fuels, which now make up 80% of our total energy supplies, for two existential reasons. The first is that our use of these carbon fuels is changing the climate and moving it out of the Goldilocks Zone in which human civilization finally blossomed. The climate outside of this zone are regimes in which a severe and rapidly changing climate presents us with a very tough fight for even the most basic level of survival.

The second aspect is the fact that the richness of fossil fuel reserves is declining and the point at which most reserves will become energetically unviable will likely be reached by 2100. Crude oil may largely be depleted decades before then, with natural gas lasting a few decades longer and perhaps the lowest grades of coal for several decades more. However, although oil, natural gas and coal are all classified as "fossil fuels", they are not directly substitutable for each other.

Major changes are therefore very much in the cards for humanity and it is our choice whether the transition will be made by our own hand. The alternative is that Mother Nature's method and timing decides how smoothly and successfully we enter the post-carbon era. At the end of the process, when we have settled into a sustainable balance, what will the level of prosperity be?

Will the future be one of bare survival for a small number of scattered outposts, or will it feature societies prosperous enough to have retained the learning we've accumulated over human history and with the capacity to continue adding more? Mother Nature will guarantee the former but only we can assure the latter.

Lots of Time to Prepare

That kind of stark choice will not come as any surprise to anyone who has paid attention to well-researched warnings that our dynamic world was going to serve up challenges from the directions of climate change and the transition to renewable energy.

We have had decades of energetically waved red flags and now the time remaining to make proactive changes can be measured in only a few short decades. Maybe a threat on the horizon two decades away is still too remote and explains our sloth-like response to date. Maybe the gradual and multi-faceted nature of the impacts from these two unfolding trends has not presented humanity with a sufficiently loud alarm for us to fully register that the danger they represent is both clear and present.

Covid-19 Sets the Response Benchmark

Such has not been the case with the Covid-19 pandemic. Once hospital hallways in Italy began filling up with dying patients, and refrigerator trucks were needed to accommodate the overflow of body bags from morgues, governments of all political stripes took notice and implemented urgent measures to shield their populations from such impacts. The Covid-19 threat was a very easy problem to visualize.

Results varied greatly but at least practically every government began to deal directly with the problem. In many cases, they found themselves poorly prepared. The streamlined globalized commercial economic system of minimal inventory and just-in-time delivery from centralized manufacturing facilities was completely the wrong model to deal with the biophysical challenges of a global pandemic.

Although the pandemic arose very quickly, we should not have been surprised that it occurred. Epidemics and pandemics have been a recurring feature of human existence since we first began to spread around the world and live in densely packed villages. To clarify, epidemics are regional outbreaks whereas pandemics spread across nations and continents.

These events are sporadic and their timing is random but, for the most part, they have tended to be regional with several spectacular exceptions e.g. The Black Death (Plague) and the Spanish Flu. Climate fluctuations are somewhat random and regional too, but major climate change, on the other hand, typically follows an overall trend that is now easily measured.

Despite this, how its complex impacts play out will be as uncertain as they are large. A decline in climate stability and an increase in extreme weather events are difficult to quantify even using the biophysical economic metrics of real physical units. Attempting to quantify the impacts in the language of the commercial economy, "dollars", is futile.

Epidemics like Ebola, Sars-1 and MERS, happen relatively often and occasionally grow into pandemics. Despite those events, few people could name a pandemic that has occurred since the misnamed Spanish flu of 1918–1920. (The label "Kansas Flu" would have been more accurate.)

Surely we should be keeping a keen eye out for events which spread around the world and kill tens of millions of people but who among us, before 2020, has ever considered a pandemic to be a prominent threat requiring enduring and costly preparation? Well, some experts have and one of them was Bill Gates. At a security conference In Munich in 2017 he warned of the pandemic threat.

Forbes Magazine[1] quoted Gates as saying: "Whether it occurs by a quirk of nature or at the hand of a terrorist, epidemiologists say a fast-moving airborne pathogen could kill more than 30 million people in less than a year. And they say there is a reasonable probability the world will experience such an outbreak in the next 10–15 years."

In hindsight, Gates' comment, made just 3 years before the Covid-19 pandemic re-arranged humanity's priorities, was about as precise and prescient as predictions get.

Note this warning was given at a security conference, not at a health meeting. It could just as well have been delivered at an economic or environmental forum because the effects of a serious pandemic are that wide and profound. Gates could have focused on some higher profile issue such as nuclear weapons or climate change but, for good reason, he chose to focus on infectious disease threats.

Our society was in need of a wake-up call and slap in the face. In response to Gates' alarm we all hit the snooze-button. However, when it arrived, Covid-19 was a rude awakening and the shock for which we should have been preparing.

Predicting that a pandemic will occur is a reliable way to be proven right because they have occurred throughout our history and will continue to do so. In just the past 100 years, three influenza pandemics occurred at intervals of several decades. The "Spanish Flu"is estimated to have caused 20–50 million deaths and both the Asian flu in 1957–1958 and the "Hong Kong Flu"in 1968 were estimated to have caused 1– four million deaths each.

Why Us?

Humanity has grown into the richest food source on the planet for some types of organisms. The earth, with its complex biosystems, is a pathogen factory and human society, being now so numerous, densely packed and closely connected, resembles something of a gourmet Petri dish. A virus with high transmissibility and very low initial symptoms can spread around the world in days before being detected. Dense urban cores with poorly ventilated living, transit and working conditions combined with high viral-emitting sporting, religious and protest events guarantee transmission and distribution.

---

[1] https://www.forbes.com/sites/brucelee/2017/02/19/bill-gates-warns-of-epidemic-that-will-kill-over-30-million-people/?sh=1c55d26b282f

Pandemics are but one of the many types of events which have led to large numbers of deaths and societal setbacks. The earth is a dynamic planet with shifting weather patterns and peaks and valleys of biological flows. Humanity has clung to the edges of existence for over 200,000 years and it is merely in the past 400 generations that we have been able to establish a reliably firm foothold across the globe.

What Event Turned Us Loose?

In the hundreds of thousands of years prior to the Goldilocks period of climate stability, which began 12,000 years ago, the earth's climate more closely resembled a stamping mill of climate fluctuation. This pattern presented extreme survival challenges to hunter-gatherer groups forcing them to adapt and move often on relatively short notice. For the past 12,000 years our climate has not been too cold or too hot to prevent sophisticated societies from taking root (Fig. 1.1).

Certainly, over the past 200,000 years, there must have been many attempts by modern humans at nascent agriculture. But even given sufficiently stable weather patterns of a few years or decades, none of these attempts was able to endure until the rare opportunity of our currently exceptionally stable climate period began 12 millennia ago.

This period allowed our ancestors to increase the areas in which they could plant and protect their crops until harvest time consistently year after year. They didn't have to abandon the progress they'd made as a result of growing conditions becoming impossible. They could stay in one area, increase in numbers, build small towns and hand down learning from one generation to

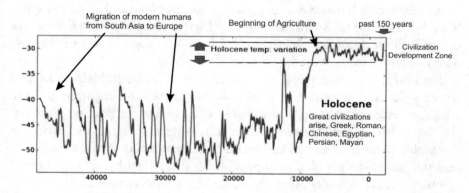

Fig. 1.1 Climate history and the Goldilocks period

From 2.2 million years ago   to 12,000 years ago to   Today

Fig. 1.2 Rapid development in the stable climate era

another. Slowly, after hundreds of thousands of years of almost no progress, human knowledge began to accumulate and accelerate (Fig. 1.2).

At that point, we could begin to avoid some of the worst impacts from what had always been assumed to be a fickle and often vengeful nature. It took humans two million years in extremely challenging climatic conditions to progress from broken stone spear points to finely crafted arrowheads. In a period of exceptionally favourable climate, it took only 12,000 years to progress from stone arrowheads to walking on the moon.

Human Population Cycles: It Hasn't Been a Straight Line Upward

Even then, nature's fluctuations have led to very substantial regional declines in human populations. Of course, under favourable conditions humans can propagate rapidly but we can also die off even more rapidly when climate and agricultural conditions turn against us. The graphs below illustrate the cyclical nature of human numbers with booms and busts seeming to occur over 300–500 year time spans.

In these cycles, humans encounter a rich resource base and begin to exploit it during a period of favourable climate. Their numbers grow and soon the society is both degrading the richest land while turning to less productive, marginal land: "moving up the hillsides" so to speak.

The now much larger population is just able to maintain itself on its land with a great deal of extra effort, possibly in the form of irrigation and soil retention systems like terraces or a change of crops and fertilization practices. But then the regional climate throws in a wiggle, bringing the period of favourable weather to an end and making the marginal existence of a significant portion of the population impossible for a number of years or decades.

Famine, social upheaval, plagues, wars and migration ensue, and the population declines dramatically leaving the land to begin to heal to a certain

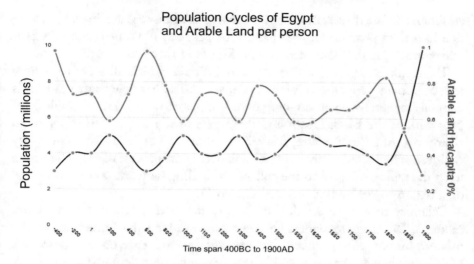

Fig. 1.3 Given the stability the Nile brought to Egyptian agriculture, its population cycles are extremely well defined

degree. Farming is once again centred on the most productive lands and the society renews itself and starts growing again.

This cycle has been repeated hundreds of times throughout human history in all areas of the world. Changing the climate, however, makes this a global issue and rather than a cycle; it might be a one-time event. The "Grow, Deplete, Collapse" pattern will occur but the "Repeat" element may well not happen as the level of climate stability human societies requires may never occur again, or, if it does, possibly not for thousands or even hundreds of thousands of years.

Egypt has experienced classic population cycle patterns and population with arable land per capita, as shown in Fig. 1.3. Arable land area is essentially fixed along the banks of the Nile River. From 1900 onward, Egypt's population has skyrocketed upward, mirroring the experience across the globe. But rather than indicating a change of pattern to infinite growth, it likely portends the start of a much larger and severe population cycle.

Human History in 500 Words or Less

Typically, humans encounter favourable food and shelter conditions and populations expand for decades and often centuries. However, at some point, our numbers grow to a level which strains the soil and hunting limits of the region, and then climate throws in a wrinkle for several or more years. Previously very rich soils then produce far less than in the past and more marginal soils produce nothing at all. Herds decline and fisheries fade away.

The result is hunger, famine, disease and strife which all serve to rapidly reduce human populations, often by 30% to 50%, in a few short years. This happens regionally on an ongoing basis but planet-wide occurrences do happen; witness the Black Death and the events of the seventeenth century, so clearly described by Geoffrey Parker in "Global Crisis".[2] This exceptionally well researched and lucidly written book details the process of societal decline from the failure of crops to the collapse of trading networks as they occurred on a region by region basis around the world.

Calamity drives a great deal of history but Parker lays out what drives calamity. We need to understand this process given the immensity of the modern human enterprise and our near total dependence on energy sources which simply will not be as readily available in coming decades.

With their gradual increase in knowledge, human populations, despite occasional setbacks, grew but slowly. However, once we began to apply fossil fuels to do work for us, our numbers exploded. Fossil fuels substantially freed humanity from many of Nature's limitations and variability, and we propagated like Covid-19 in a crowded karaoke bar (Fig. 1.4).

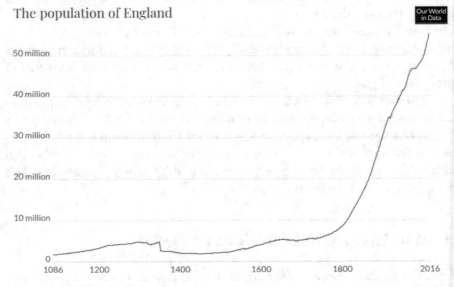

Fig. 1.4 Population of England 1086–2016

[2] https://yalebooks.yale.edu/book/9780300219364/global-crisis

As can be seen from the above graph of population growth in England, since fossil fuels became readily available, humanity has increasingly been able to ignore nature and her rules. By harnessing the vast, and, until now, seemingly unlimited, stores of energy, humanity has exerted its dominion over natural processes. But our small victories over nature are Pyrrhic in nature and invite a final defeat in the war as nature begins to play her increasingly more powerful trump cards of climate change and resource depletion.

For those paying increasing attention to the natural world, the assumption of endless growth in population and material consumption appears to be reckless and quite possibly suicidal. Yet, the mantra of the commercial economy is one of business-as-usual and relentlessly increasing consumption. Real-science based biophysical economics is actually a study of the business of survival and long term prosperity and it assumes quite the opposite. Thus the physical sciences find that our society must be brought into line with natural limits and tuned for human wellbeing.

A Blade-Runner type, dystopian existence might measure well on the GDP index, but few people would want to live in such a world. Neither would the massive urban ghettos portrayed in the film be supportable by natural systems. We have to look beyond the maximum consumption commercial model that has evolved over the past 500 years to take a different tack that restores environmental health and maintains a high degree of human well-being.

In some cases, it is best to beat back a threat and in others, it is best to simply get out of the way. We need to avoid open confrontation with Mother Nature. Living among natural flows is vastly more efficient and sustainable than erecting walls and fighting natural processes on all fronts, all of the time amid relentlessly declining resources.

The Development Template

Moderate and consistent climate gave rise to agriculture leading to stationary and growing populations which condensed into small villages and towns. This allowed labour specialization to arise, and trading networks and technology transfers to grow rapidly.

More efficient food production freed up a large proportion of workers from agricultural labour who could then engage in other pursuits. Science and innovation built on themselves yielding continuous increases in productivity and eventual improvements in social equality. Progress was slow and occurred over millennia until the widespread use of fossil fuels literally blasted human development into space.

Innovations that could be considered industrial in nature have cropped up in many different geographical locals all through history. The first machine that transformed heat energy into mechanical energy was Greek/Roman as was the stunning Antikythera Mechanism, a gear-driven celestial computer.[3] A Roman is said to have invented unbreakable glass[4]but was executed for it. The unhappy fates of those who threatened to disturb the stability of the established order is a recurring theme in human history.

Innovations, from gunpowder to the printing press, abounded in China's history. The first forms of steel were made in Anatolia (modern Turkey) in around 1800 BC and Wootz steel, a pioneering steel alloy, was made in India around 300 BC.

The timing and motivation may have been different for every nation but the steps to modernization were made up of the same building blocks.

Modern humans doubtlessly experimented with agriculture all over the globe in the 200,000 or so years before it finally took hold and grew permanent roots. Small, early attempts at agriculture may have been successful possibly for decades. But, that progress would have been wiped out by climate shifts which forced humans to move and return to a 100% nomadic hunter gatherer existence.

The exploitation of the vast tracts of virgin forests of eastern North America only began with the arrival of Europeans and their iron and steel tools. These allowed work to be done vastly more efficiently than with the stone implements of the indigenous inhabitants. The same process occurred in Britain in the early Bronze Age when the fine and durable edges of bronze axes were applied to the great virgin forests of the land. The forests were cut down and replaced by farms during a period of warm dry climate, allowing the population to expand and prosper on the rich soils of Southern Britain.

Britain

Britain was the first country to see widespread and continuous innovation and application of truly industrial (and truly reckless and environmentally damaging) processes. It was the first country to marry its enormous coal and iron ore reserves to hammer out a wide variety of goods using a rapidly evolving phalanx of industrial processes and machines. Mass production

---

[3] The Antikythera Mechanism https://www.bbc.com/reel/video/p09pcwnz/unlocking-the-secrets-of-the-world-s-oldest-computer?utm_source=taboola&utm_medium=exchange&tblci=GiB5vTs3iHW kG6CPBSrnMRhCt2TrUeHieEacoChsJ3WucCCMjFQoyZb7rdro8rRv#tblciGiB5vTs3iHWkG6C PBSrnMRhCt2TrUeHieEacoChsJ3WucCCMjFQoyZb7rdro8rRv

[4] https://en.wikipedia.org/wiki/Flexible_glass

labour saving and job specialization had finally evolved into an industrial juggernaut which was unleashed by an unwitting humanity on an unsuspecting world.

The widespread use of coal happened none too soon because English forests were dwindling due to the unsustainable demand for firewood, and, to a lesser extent, the building of fleets of ships. Sweden experienced the same process in the mid-1800s when its forests collapsed and one in seven Swedes either migrated or died.

By 1700, "England coal was already used on a large scale long before the introduction of steam engines and industrialization, simply to replace firewood for heating purposes" (Warde, 2007[5]). This makes clear the point that the forests were either gone or were rapidly depleting, a natural consequence of a large and growing population on a small island. Hence the need to employ coal as a heat source and for England to look beyond its own borders for ever more resources.

Coal, it should be said, was nothing new to England or Germany. The Romans had imported it into England from Germany almost 15 centuries before it rose to prominence as the transformative energy source it was to become. The Romans used coal for heat. When the English learned to use coal to do mechanical work via the steam engine, the Industrial Revolution was launched.

Rapidly increasing industrial capacity, with its increasing numbers of superior tools, launched the island nation and gave it the ability to form the British Empire and project its power around the world. The British had more and better ships and more and better guns. They also had a burgeoning population of abysmally poor people in urban slums who could be pressganged into service as soldiers and sailors.

Life could be short in the armed forces of the Empire but it was a lot shorter for the poor in the inner cities. Life expectancy could fall to as low as 21 years in the worst slums. Compare this to the life expectancy of 26 years for a Roman slave and between 25 and 35 years for an African slave in America.

The Industrial Revolution was truly a huge leap of progress, but it was not all pork pie and Guinness as these photos of Birmingham and a young coal miner illustrate. Children as young as six were put to work in factories for up to 14 h shifts, 6 days a week. Pollution and working conditions were appalling but they laid the groundwork for progress (Figs. 1.5 and 1.6).

The Industrial Revolution was dirty, dangerous, often inhuman and, given the complete dependency on coal, ultimately unsustainable. But it was a societal launch event for humanity that propelled us past our agrarian limits. We were propelled into the uncharted world of science, consumerism, unrestricted resource exploitation and a complex and interdependent economy.

---

[5] Ward, 2007, Energy Consumption in England and Wales 1560–2000.

Fig. 1.5 Early Industrial development came with a heavy environmental price

Britain's powerful engine of empire was used to establish control over resources around the world and ship them back to the small, but now mighty, island kingdom. As Dr. David MacKay said in his landmark TED Talk,[6] "A Reality Check on Renewables", "the immense reserves of coal which contained as much energy in them as all the oil under Saudi Arabia, and it powered the industrial revolution and put the "great" in Great Britain."

Immense learning and progress took place and a transition to other forms of energy (oil and natural gas, electric and nuclear) occurred before the rocket fuel of coal petered out in Britain and, to a large degree, in Germany. China still runs on coal but their reserves will only last several more decades. This explains their emphasis on importing coal and other forms of energy.

British coal production peaked in 1918 and then declined and, by 2000, production of oil and gas from the North Sea had also peaked. Once the hub of industrial might and innovation, Britain now imports most of its manufactured goods and energy and is a net food importer as well.

Germany and Northern Europe

The development of Germany and northern Europe followed closely on the heels of Great Britain with innovations and industrial capacity leapfrogging each other. Germany also had huge reserves of coal to go along with its rich

---

[6] https://www.ted.com/talks/david_mackay_a_reality_check_on_renewables?language=en#t-18991

Fig. 1.6 The human toll was steep as well. This is John Davis (1897–1963), after his first day working down in the pit as a miner in Wales at the age of 12

forests and fertile soils. It had its own iron ore reserves and ready access to those of Sweden. Combined, this meant that growth was unconstrained.

Germany did not have the vast overseas empire that Britain did, but nevertheless prosperity, population and general welfare increased greatly through the 1800s. However, the lack of an empire, with its guaranteed ready access to critical resources, chief among them oil, led to tensions which repeatedly boiled over into wars. The theme of a nation's leadership having ambitions realizable only by using foreign resource bases is oft repeated throughout human history.

USA

In the colonies, development time was extremely compressed since the settlers arrived with a full toolkit of all the technology developed in Europe over the previous thousands of years. Settlers could hit the ground running and quickly build up the same level of productive and transportation infrastructure they had left behind in their mother countries. In fact, replicating the pattern of life in the mother country was often carried out to a fault as the dictum "Make the world England" implies.

Not only did the United States of America have access to the most advanced technology of its time, it also had essentially unlimited and virgin stores of resources to which to apply it. Farmland, timber, coal, minerals, then oil; the United States simply had no limits for the first few hundred years of its development. For a few centuries, "endless growth" really did appear to be just that.

The American population grew rapidly and the improvement in living standards and individual prosperity grew right along with it well into the 1960s. After that, increasing limits in the form of oil shortages, pollution crises and quality of life issues began to be felt.

Canada

Given its extreme climate, the first settlers needed huge amounts of energy to heat their leaky log homes and buildings with very inefficient open hearths. After all, the average January nighttime temperature in Quebec City was 18 °C (or 32 °F) colder than that of Paris or London. Fortunately, settlements in Eastern Canada were essentially cut out of virgin stands of hardwood so, despite huge heat demand, the needed firewood was readily available.

As farmland displaced forest to a large extent, coal had begun to become available. Like the United States, development in Canada occurred over a few short centuries whereas societies in Germany, England and China took many thousands of years to evolve from late stone-age to industrial societies. Also, per capita demand for energy was higher than virtually any other country and remarkably, stayed approximately level from first colonization to full industrialization.

The types of energy input changed with the firewood yielding to oil and electricity. Most importantly, these new fuels allowed a wide variety of uses and much greater efficiency. For 400 years, from the arrival of the first Europeans to the 1950s when oil really made its impact, per capita daily energy consumption had remained roughly level in Canada.

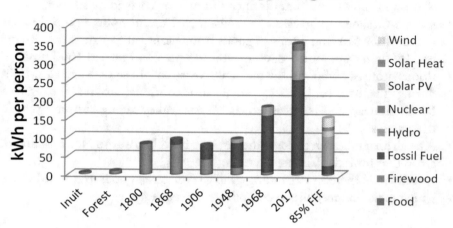

Fig. 1.7 Energy Budgets from Inuit, forest Amerindians to settlers to the consumer era and eventually 85% fossil fuel free

Where once 100% of the energy used was for heating and cooking, people of the twentieth century could travel across the country, do vastly more physical work, and produce a wide array of material goods on the same energy budget that was simply required to keep the first settlers from freezing to death (Fig. 1.7).

When oil launched the personal transportation car fleet in the 1950s, energy consumption soared and it jumped again when natural gas became available to heat much larger houses. Oil could be used for many more applications than coal; chief among them being the icon of the modern consumer era, the automobile, in addition to the newest form of luxury transport, the airplane.

In contrast, the energy budget in the Fossil Fuel Free era (FFF) will be much lower than it has been for the past two generations.

Consider the changes in health, security, comfort, learning and the quality of life that have sprung from the chain of developments enabled by ready access to virtually unlimited energy.

A short list of changes wrought by fossil fuels

• The hours of labour required to heat one's home was possibly 150 h of woodcutting versus 30 h with a chainsaw and no time at all if heat was supplied by electricity or fossil fuel.

- Personal travel meant a day's walk of 40 km in the years prior to the arrival of the automobile but that distance can now be covered in 20 min.
- A cross-Canada trip by canoe took five men 4 months plus a possible overwinter while carrying 3 tons of goods from Montreal to Vancouver. By the late 1800s, a train took 5 days, maybe 3–5 men and carried 1000 tons.
- Roman tourists who visited the 1500 year old house of Menelaus, built during the age of Ulysses, might also continue around the eastern Mediterranean and have visited the Aswan Valley taking almost a year in the process.
- A jet airliner takes 5 h to cover the distance from Norway to Newfoundland, a trip that took the Vikings 8 weeks.
- And there is the ever expanding galaxy of material goods and innovations which make modern life so complicated, yet so easy.

Japan

In the 1850s, American gunboat diplomacy forced Japan to open itself to trade which caused the Japanese government to adopt drastic reforms to preserve their culture. They abolished the feudal system and modernized their military as well as laying the foundation for rapid industrialization. Strong support for technological and industrial development proved its effectiveness when, in 1905, the Japanese decisively defeated the Russian Imperial fleet in the battle of Tsushima. Predictably, they were fighting over control of someone else's resources; those of Korea and Manchuria.

Thirty years later, Japan found itself in the same position as Germany since domestic ambitions greatly exceeded what their domestic resource base could support. Again, in the Second World War, Japan was sufficiently desperate for the fruits of empire that it was willing to directly attack the two largest military powers of the time, the British Commonwealth/Empire and the United States. Although the war was an extended one, and despite impressive early victories by the Axis powers, their defeat was largely inevitable.

A smaller, resource-short nation attempting to knock much larger resource-rich nations out of a war with quick victories seemed to be a theme of the first half of the twentieth century. But when dealing with large resilient foes, "quick" doesn't necessarily translate into "decisive", as both Japan and Germany were to learn.

Following the Second World War, Japan rebuilt itself into one of the most technologically and socially advanced countries in the world with very high levels of equality, health and safety.

Russia

In a similar way, Russia, which suffered when the western powers intervened in the Russian Civil War in the 1920s, adopted rapid industrialization to assure national survival. The Soviet Union's centrally controlled economy decided to heavily invest its resources to enhance its industrial production and infrastructure. It was able to do this just in time to defeat the German invasion of WWII and thereafter quickly evolved into a superpower complete with extra-territorial ambitions of its own.

In 1957, the Soviet Union shocked the supposedly more technologically advanced western nations by launching the first satellite, Sputnik 1, into orbit.

China

Even before civilization began to take root in China some 5000 years ago, people in the region were already making use of coal. The incentive to become more dependent on coal grew as forests declined and population increased. In the period from 1700 to 1950, forest decline in some regions ranged from 20% to 50% while population increased from 200 million to 550 million.

After centuries of foreign domination and societal chaos, China, under Mao Zedong's Communists, established a level of stability. The first attempt at rapid industrialization, "the Great Leap Forward" was a disaster but subsequent leadership has found the formula for success and China is now a dominant industrial, and increasingly, social leader.

Like all large and powerful countries, China has outgrown its domestic resource base and has successfully implemented a program of foreign investment to establish ownership of key resources and transportation infrastructure. This neo-imperialism has so far been accomplished without the use of military force.

Modern Economies developed in different times and different places but the flowers of industrialization all needed the base of climate stability, the seeds of resources and the sunshine of fossil fuel energy to finally bloom.

Table 1.1 Coal dependency date

England[a]	1705
Germany	1863
France	1865
Netherlands	1878
USA	1885
Canada	1906
Sweden	1927
Italy	1929
Russia	1930
Spain	1942
Portugal	1966

[a]England reached a 90% coal dependency level in 1844

Fossil Fuel Transition Timeframe

Energy availability drove societies forward, in some cases whether they liked it or not. "It is not environmental determinism to conclude that changes in energy sources have dramatically shaped the choices people make as a society and that those choices come with material, as well as political, social cultural and economic opportunities, constraints and consequences." R. W. Sandwell "Powering Up Canada the History of Power, Fuel and Energy from 1600".

Below are the dates when countries stepped beyond wood and reached 50% of their energy supply from fossil fuel and electricity (Table 1.1).

Overall, per capita daily energy use in Europe averaged 23 kWh per day in the period 1500–1850 but, once the fossil fuels dominated the picture, consumption increased by almost a factor of 5 to 106 kWh/day.

Clearly this huge increase in energy consumption pushed billions of people higher up the scale of "needs" and into the realm of "wants". When large injections of fossil fuels hit the bloodstream of human society, huge numbers of people were launched several rungs up the Hierarchy of Needs ladder. They did less work, produced more and had more free time and more options.

How can we best anchor ourselves high up on the pyramid if the energy and environmental systems we depend on begins to weaken? (Fig. 1.8).

Maslow's Hierarchy of Needs graphic depicts the transition from needs to wants as well as the building blocks of a cohesive society.

Impact of Fossil Fuels

Fossil fuels gave humanity the ability to live outside the constraints of natural energy flows. We could survive where we had never been able to and grow more crops more consistently than ever before. Our numbers increased

Fig. 1.8 Needs are followed by wants

dramatically, and they increased to a much larger degree in the cold and challenging northern regions than in the much more amenable temperate and tropical regions.

In the year 1500 AD, just as Europeans were beginning to arrive in the Americas, the population of Inuit in Canada's Arctic was around 2000 people. That is 2000 people in an area of 1.4 million square kilometers. The population of Europe at the time was around 90 million in an area 1/3 smaller.

Canada's population in 1500 AD may have been 250,000–500,000 with the majority located in the moderate climate of British Columbia's west coast. The limits set by extreme climate and geography constrained the population density to an average of one person per 20 km² over Canada's area of ten million square kilometer.

Compare this to the population density of the resource rich and climate moderate Central America where the pre-contact density of Mayans and Incas could reach 100 people per square kilometer. This is 2000 times more dense than in Canada as a whole and 70,000 times more dense than in the Arctic. Latitude and climate matter.

But fossil fuels changed the math. In 2021, with a population of 38 million, Canada's population density had grown to 3.8 people per square kilometer, while Mexico's population has increased proportionately far less. Having almost endless energy on tap is a great leveler of nature's extremes.

What happens when that great store of fossil fuels is no longer available is the subject of later chapters.

Table 1.2 The fossil fuel / population / latitude multiplier effect

Country	Year 1500 (millions of people)	Year 2020 (millions of people)	Ratio 2020 versus 1500
Germany	12	83	7: 1
Japan	17	126	7: 1
Mexico	10	127	13: 1
Britain	3	67	22: 1
China	60	1400	23: 1
Russia	6	144	24: 1
Australia	0.75	25	33: 1
United States	6	328	54: 1
Canada	0.5	38	76: 1

Population Comparison and the possible fossil fuel multiplier effect (Table 1.2).

Populations in pre-fossil fuel societies varied greatly due to plagues, wars and famine. Over any 50 year period there could be a 30–40% difference.

The "new lands" of Canada and Australia have low population densities due to the harshness of geography and climate over most of their areas. Relatively small portions of their total areas are suitable for agriculture or amenable to human habitation. Similarly, there are many areas on the Eurasian and African continents which have been subject to human occupation for hundreds of thousands of years yet still have low population density for the same reasons.

Kazakhstan has a population density of seven people per square kilometer, Mongolia 2, Namibia 3 and Libya 4. These low densities speak to the challenge of their climate and the low potential of their resource bases.

The final contribution of fossil fuels is, like that of any other good leading social contributor, in the development of their replacement. Electricity is the heir apparent and, realistically, the heir inevitable. Accompanied by continuous learning and increasing wisdom, the transition to renewable energy, holds out the possibility that our future might be better than our past as we move into balance with a healthy Earth.

A period of exceptional climate stability allowed humans to take control of their food supply but energy is the steed that carried humanity into the modern age. Fossil fuels are nearly exhausted from their 400 year run and humanity now finds that it is necessary to begin to change horses.

2

Energy and Our Lives

Basic Building Blocks

Now that we've arrived as a planet-dominating species, it's important to understand the underpinnings of our complex society. After biosphere and climate, energy is the most critical element of our existence. Initially the only energy we controlled was that embedded in our food supply. It allowed bare survival. We then learned how to control fire, first by taking it with us when needed and then by being able to create it wherever we went.

Once we had added fire to our toolkit, we set about developing better fuels for different purposes. Animal fat, seal and whale oil for light and some heat; wood for heat and cooking and animal dung and peat when the more energy dense wood was not available.

This level of energy competency was enough to allow us to begin to work with metals and create tools vastly superior to the stone and wooden implements which had been evolving so slowly over hundreds of thousands of years. The extraction and refining of metals required much higher temperatures than those used for cooking food and charcoal was able to deliver the required heat. But charcoal was made from wood. It required considerable work to produce and had distinct limits on the maximum heat it could generate.

The time and effort to fell a tree, cut it into suitable lengths and split it using steel tools as considerable. This was energetically impossible using stone tools. The wood then had to be stacked for drying over a period of many months depending on weather conditions. Firewood also required substantial areas for storage and it was subject to degrading rot if not used in a timely fashion.

J. E. Meyer, *The Post-Pandemic World*, https://doi.org/10.1007/978-3-030-91782-1_2

Coal did away with all of those constraints. It could be burned the moment it was hauled out of the ground and offered both greater energy density and higher heat potential than any wood based process. Coal could be readily broken up and shipped in sacks and barrels and could be left outside for years or decades without losing any of its potency.

In the time, maybe 3 days, it took a man with an axe and crosscut saw to fell, cut up and split a cord of wood with an energy content of perhaps 7300 kWh, a coal miner could shovel perhaps 15 tonnes of coal with an energy content of 125,000 kWh. While the woodsman had to wait months to burn the fruits of his labour, the miner could take home enough coal in a bucket to keep the house warm each night.

Three hundred years ago, a bush cord of wood (approximately 4 m^3) in a cabin with a stone hearth in a northern region might provide comfort for 2 weeks during the winter.[1] One hundred and fifty years later, a tonne of coal burned in a moderately efficient coal stove could easily heat the same cabin for a month or more. Consequently, noxious clouds of coal smoke began to pollute the towns and cities of the Industrial Revolution.

Despite the unhealthy aspects of fossil fuels, they proved to be a far superior means for producing heat, particularly as forests were being progressively diminished by centuries of expanding and unsustainable exploitation. Regions with coal reserves prospered.

Fossil fuels began their service to humanity by providing heat. Not just heat, but heat anytime and anywhere because, aside from their tremendous energy density, fossil fuels have another characteristic which makes them so ideal: they are self-storing. Neither an electric battery nor any high tech apparatus is required to preserve their energy, as they do this themselves. Fossil fuels are heat batteries. So is a stick of wood but it just doesn't have the energy density or the ease of handling that fossil fuels do.

If you have a can, you can store gasoline. Natural gas can be stored in low pressure tanks and coal can be stored in a box or just left piled up in a corner of the basement. Fossil fuels were easy to mine, easy to store and very simple to use. Take one portion of fossil fuel, add a match, and heat and light are produced instantly. The ability to produce heat and light on demand was a huge step forward.

But then the major technological breakthrough occurred. Humans found a way to do mechanical work by controlling the heat that was the primary

---

[1] Heat content maple = 25 million BTU, coal = 28 million per tonne, 15 tonnes = 420 million BTU
25 million BTU = 7300 kWh, 420 million BTU = 125,000 kWh.

output of fossil fuel combustion. This controlled energy could then be applied to a growing galaxy of tasks.

Tools plus energy made humans vastly more efficient and able do types and amounts of work never before dreamed of in our world of stone and wooden implements. Our tools allowed us to use more energy more effectively and efficiently.

The physical work an average person can do unaided compared to what we can do today courtesy of mined energy is well laid out in the book The Energy of Slaves, by Andrew Nikiforuk. By 1824, steam power was doing work equivalent to 750,000 men and coal effectively added the work of three billion slaves by the late 1800s.

In terms of direct energy equivalents, earth scientist David Hughes has estimated that a healthy individual on a bike riding an 8 h day with normal holidays would take over 7 years to output the energy contained in 1 barrel of oil.

Clearly, we fire apes have come a long way in terms of technological sophistication. In terms of social and environmental wisdom though, we are lagging somewhat. Currently, we are more clever than we are wise. The complexity of the human enterprise is such that it has both created existential threats to humanity as well as the rest of the riders on our planetary bus while making those same problems very difficult to solve.

Energy Is the Base Currency for Both Human Society and Nature: How Do We Measure It?

All life is completely dependent on energy. No energy, no movement, no life. The amount of energy we are able to harvest or control, combined with the efficiency with which we are able to use it, determines how numerous, prosperous and resilient we are.

In this book I will be representing all energy measurements in the metric typically used for electricity production and consumption which is kilo Watt hours. In plain language 1 kWh is the amount of energy consumed if a rate of energy consumption of 1000 W occurred for 1 h. Turn your 1 kilo Watt electric heater on for 1 h and you will have consumed 1 kWh of electrical energy by turning it into heat.

A Watt is a unit of power – the ability to do work – or the rate of energy transfer. Scientists would vastly prefer to use joules (a Watt is 1 joule per second) but people are used to seeing their electricity bills in kilo Watt Hours (kWh) and, increasingly, they are "filling up" their cars with electricity in kWh. Accordingly, this book will use kilo Watt hours as its base energy yardstick.

The renewable energy future is electric and both generation and consumption will be measured in kWh. As we'll see, electricity won't be the only form of energy we'll be working with but it will be the dominant one.

A kilo Watt hour in Perspective: Consumption

Few of us understand how much energy we use for any particular purpose. But knowing where you use energy and how much you use is important for your plans to conserve, produce and store energy. Below is a table of everyday devices which use electricity to do a number of typical tasks.

It is easy to see the scale of the energy used and the end cost based on a moderate[2] electricity cost of 20 cents per kilo Watt hour which includes all taxes, delivery and service charges (Table 2.1).

The costs of driving an electric vehicle compared to an internal combustion engine (ICE) vehicle will be discussed in some detail in a later chapter.

For a comprehensive and sober look at renewable energy and our energy demand, download the free book "Sustainable Energy – without the hot air" by David MacKay, www.withouthotair.com. It is a tremendous resource.

Energy Production (Table 2.2)

Unlike the fossil fuel world, in which combustion-generated heat is used to drive all manner of processes, electricity can be produced by a number of methods. Any mechanical motion such as falling water, blowing wind, turning crankshafts, can be used to turn an electrical generator. Nuclear plants take the heat energy released from splitting atoms and turn water into steam, which is then used to turn turbine generators.

But heat can also be derived from both the sun and possibly from geothermal sources. Heat is a lower grade form of energy meaning that it is easier to harvest but it is also limited in its density, the work it can do and how far it

---

[2] Electricity costs per kWh range from 6 cents US in Russia to 38 cents US in Germany. The cost varies greatly by region in each country and is dependent on the source. Hydro and coal are typically cheapest and nuclear the most expensive.

Table 2.1 Energy consumption and cost for everyday uses

Item	Power rating (draw)	Length of use	Notes	Math	Cost @ 20 cents/ kWh
Plug-in heater	1000 W	1 h	1.0 h	1.0 × 1000 = 1000 Wh or 1 kWh	20 cents
Kettle – to boil a 500 cc cup of coffee	1500 W	3 min	1/20th (0.05) of an hour	1500 × 0.05 = 75 Wh or 0.075 kWh	1.5 cents
Hair dryer	900 W	10 min	1/6th (0.10) of an hour	900 × 0.1 = 90 Wh or 0.90 kWh	1.8 cents
Plug-in circulation fan	60 W	12 h	Running all day	60 × 12 = 720 Wh or 0.72 kWh	14.4 cents
Hand circular saw	10 amps at 120 V = 1200 W	20 cuts at 10 s each = 200 s	200 s = 3.3 min or 0.055 h	1200 Wh × 0.055 = 0.067 kWh	1.3 cents
Quick shower	20 kW on-demand electric tankless water heater	6 min	6 min = 1/10th of an hour	20,000 Wh × 0.10 = 2000 Wh or 2 kWh	40 cents
Electric car compact	0.14 kWh/km	100 km	Measuring energy use over distance not time	0.14 kWh/100 km =14.0 kWh	$2.80 2.8 cents/ km
Electric car large	0.4 kWh/km	100 km	Measuring energy use over distance not time	0.4 kWh/100 km =40 kWh	$8.00 8 cents/ km
Compact gas powered car	7 L of gas per 100 km (34 mpg US) (40 mpg Imperial)	100 km	Measuring energy use over distance	7 L/100 km =7 L	7 L @ $1.20/L = $8.40 or 8.4 cents/ km[a]

[a] 7 L equals 1.8 US gallons or 1.5 Imperial gallons and the cost / mile would be 13.5 cents

can be transported. Solar hot water panels can be 50% efficient whereas their electrical PV counterparts manage less than half that.

The high grade energy, electricity, can be transported hundreds of kilometers down a cable and lose less than 10% of its energy whereas the low grade

Table 2.2 Energy production comparison

Source	Rate of production	Daily production	Annual production	Annual energy output	Notes
Bicycle generator	0.2 kW/h	30 min	183 h	37 kWh	Cycling at home at a very high level of exertion for ½ h every day for a year would provide enough energy to heat an efficient small home for 1 day in the winter
Single solar photo voltaic panel	0.3 kW/h	6 h	2190 h	657 kWh	6 h or ¼ day of strong sun for a 25% capacity factor equal to a location in the US South West
Solar farm in the UK 100 acre	20,000 kW/h	6 h	2190 h	44 million kWh	25 acres of land is required for every 5 megawatts (MW) of installation
Wind turbine 2 mW	2000 kW/h	12 h	4380 h	9 million kWh	A large turbine with 12 h of strong wind per day gives a highly optimistic capacity factor of 50%
Nuclear generator (CANDU)	900,000 kW/h	24 h	8760 h	8 billion kWh	This assumes no down time and maximum output all of the time meaning a capacity factor of 100%
Solar hot water panel (same size as PV panel)	0.6 kW of heat/h	6 h	2190 h	1314 kWh	Heat is low grade energy and easier to collect than high grade electricity so capacity factor is twice as high as the PV panel

energy hot water[3] can only be piped for short distances before the heat loss becomes unacceptable. Low density hot air can typically be distributed over even shorter distances, probably not exceeding 20 m or 70 feet. This is why some large houses have two furnaces but only one electrical panel.

Energy Consumption by Country

Note that the energy consumption levels below include not just end use consumption, but energy used in energy production, in other words all facets of a modern society.

What they don't include is the embedded energy in imported goods. Currently a good part of our manufactured goods are made in China and elsewhere where the primary energy source is cheap and dirty coal. More on this in a later chapter, but the consumer goods you buy take a lot of energy to produce. So do the foodstuffs that are trucked and flown in from other countries.

Once again, the effect of climate and latitude on energy consumption is clear (Fig. 2.1). Northern nations use more energy for building heating and

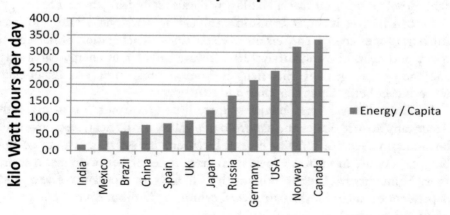

Fig. 2.1 Energy budgets for a range of countries

---

[3] The high water mark for hot water transmission would be a possible pipeline running from Iceland 700 km to the Faroe Islands. The exceptional heat potential of Iceland's thermal springs might make this technically possible due to their extreme heat and high volume capacity.

transport. In nations with more moderate climates, less heat is required for fewer months a year and more sun is available for more months a year. Typically, people are able to walk and bike safely and comfortably for perhaps the entire year and local produce is available within the active transportation radius.

Energy Budgets

What is your daily energy budget? You use a variety of energy for your vehicles, tools and home. With natural gas, oil, propane, electricity, gasoline and firewood usage spread over a year and probably varying by season, calculating your total consumption and developing a budget can be pretty complex.

But Fig. 2.1 lays out the averages per person per day for a number of different countries. This reveals the scale of your consumption. It indicates the measures needed by individuals and governments to make transition from dependency on fossil fuels to an almost complete reliance on renewable energy. "Almost" is an important caveat because, if we dramatically reduce our fossil fuel usage, we can both hit our climate targets and extend the viable life of fossil fuel reserves by centuries.

A cold turkey 100% changeover to renewables featuring zero fossil fuel consumption simply isn't a viable option for any country. But, given a strong start and consistent effort over several decades, we can come close enough to achieve both climate and social stability while also stabilizing the cost of energy.

Once a budget is set, it is possible to develop a rational strategy for the conservation of energy, conversion to electricity and generation of both electricity and heat. Life and survival have always been about energy harvesting and energy management. Electricity is the next rung up on our ascendency towards ever better energy supplies and management.

Until we discover what energy or force little green men use to toy with American Airforce planes and zap from one galaxy to the next, electricity will be our main means of applying energy. Electricity is a superior form of energy but that doesn't mean fossil fuels don't have superior performance in some areas. Humanity has been working with fossil fuels for thousands of years and dependent on them for the past several centuries. Without a word of a boast, we've picked up a few tricks along the way.

Electricity Takes Over

But, when we have 200 years of reliance on electricity under our belts, we may well see fossil fuels as being as anachronistic as seal oil lamps. They will still work but we'll have much better processes and tools to make them irrelevant. We hope.

Electricity can be made to do the vast majority of jobs fossil fuels currently do but there are some functions where electron flow cannot replace combustion. Given time, electricity will take over completely and it will offer vastly higher efficiency (no combustion process remember, just work done) which means that our overall gross energy consumption will drop significantly.

No Going Back

If a snapshot were taken of the world in 2021, it would show some problems and it would also show humans in control of many of the earth's biosystems. But does this represent control of a steady state, sustainable system or does it represent the mining of declining resources? Trends are more important than snapshots and according to the a 2011 biomass census complied by Vaclav Smil,[4] human activity over the last 5000 years has reduced total global biomass by about 50%.

Our technology has grown at a tremendous rate but have we learned the right things and built the right tools to allow both humanity and the planet to prosper going forward into the future? What if we were confronted by an unexpected biophysical threat, how would we handle it if we couldn't simply steamroller right over it? We now have the ability to be able to generate very high grade energy from sunshine, wind and flowing water but these sources are far more diffuse and variable than the self-storing extremely dense fossil fuels on which we have built our modern world.

The Niagara Falls drainage basin covers 700,000 km² of the middle of the North American continent. The output from the water that is channeled over the falls provides the electrical energy of slightly more than one nuclear power plant with 4 Candu reactors. The nuclear plant takes up just 1 km².

The amount of electricity the power plants at Niagara Falls have the capacity to output is close to 4.9 million kW. The US plants have a capacity of

---

[4] www.vaclavsmil.com/wp-content/uploads/PDR37-4.Smil_.pgs613-636.pdf

roughly 2.7 million kW, while the Canadian side's combined capacity is close to 2.2 million kW.[5]

That's enough to power 3.8 million homes using the current energy mix. But this mix also includes fossil fuel, mostly natural gas, for heating. If only electric energy was used to provide the say 80 kWh of heat these houses would need for a winter's 24 h day, then The Falls would only be able to support 1.5 million homes.[6]

This energy-density analogy can be extended to energy production from wind and solar generators. Renewables need to cover a very large area to produce the energy we currently require.

There is no going "back to the land" when the land has been largely depleted and there is no going forward with the fossil fuel consumption of the last several centuries. Progress depends on finding a new path.

Yes, There Will Be Changes

Change may not be what many people are looking for, but change is coming and the better one is prepared for it, the less painful it will be. In fact, many of the changes outlined in this book will be beneficial in both terms of financial stability and lifestyle. And then there is the small consideration of saving the planet or at least stepping clear of the worst fallout from climate change.

Energy mindfulness will be a great asset in dealing with the changes of habit and consumption which will be necessary. For people who have paid attention, the great majority of the changes we will undertake will present opportunities for an improvement in the quality of life for themselves, their families and their communities.

With Great Power Comes Great Responsibility

Humanity has been launched by a period of exceptionally favourable climate and a once-in-a-species energy bonanza to the level of becoming masters of our own destiny as well as masters of the destiny of most of the other species on this planet. How well are we going to handle this responsibility?

By now, we've recognized that we aren't at war with the planet anymore but rather are in charge of it. We need to recognize that, as much as we are global

---

[5] Nyfalls.com
[6] Energy produced by Falls over 24 h = 4.9 million kWh × 24 divided by 80 kWh = 1,470,000.

masters, we cannot simply dictate outcomes. Cooperation with the natural world is essential.

Instead of assuming the meteoric rise of human fortunes over the past four centuries will be continued by doing the same things, we need to look forward and around us. The rear-view mirror can no longer offer good guidance. We need to recognize what is good and what is sustainable rather than trying to perpetuate brief moments from the past when it seemed that vainglorious consumption was the path to eternal progress.

It has been fun, but we need to drop out of the super-consumer club. Life members may still feel they are being cheated unless they get more for less despite having to rent storage units to accommodate the "stuff" their large houses can't hold. We need to learn how to do better with the levels of exploitation nature can sustain.

As Mike Nickerson says in his sustainability site[7] "More Fun, Less Stuff" is the best and happiest way forward despite the daily promotion of consumption and growth-dependent interests in the Main Stream Media (Fig. 2.2).

We can celebrate the high levels of personal mobility which arrived in the 1950s but re-living the '50s or demanding an ever greater level of consumption is a path best avoided.

Fig. 2.2 When oil triggered the consumer era in the 1950s, it seemed the party would never end

---

[7] http://www.sustainwellbeing.net/MFLS.html

Fortunately, we have experienced the slight ripple of the Covid-19 pandemic to gently remind us we have bigger fish to fry than maintaining the golden age of consumerism. The world needed a wake-up call and Mother Nature delivered one in the form of a full blown, modern-age pandemic. Although we've stumbled badly in dealing with it, finally understanding that we can stumble is an invaluable lesson and a great warm-up or stress-test for the biophysical threats which are swirling around our ankles.

Benjamin Franklin
By failing to prepare, we are preparing to fail.

3

Pandemic Overview

What Just Happened

Short of war, there are few crises which descend on governments as clearly defined as pandemics. They arrive with little notice and completely rearrange priorities while doing tremendous damage to the internal workings of the country (Fig. 3.1).

When confronted by the Covid-19 pandemic, different governments reacted differently in terms of time, clarity and decisiveness. Every nation had their own specific mix of challenges but the overarching threat was the same for all.

What unfolded on the world stage was a living lab of responses featuring wildly varying levels of success. Some countries appraised the threat properly and took strong and correct measures very quickly. One country implemented a clear policy early on but it appears to have been deeply flawed. The leadership of some countries initially denied there was a problem. Others acknowledged the problem but felt they either couldn't do anything about it or chose to set a level of tolerable infection rates.

In the end, all governments were forced to take the pandemic seriously and implement strong policies to protect their citizens from harm, or at least minimize the harm they suffer.

Here are the proactive approaches roughly categorized.

1a. Get on the problem early and hard. Pull out all the stops and control the spread of the virus via complete and rigidly enforced lockdowns.

© The Author(s), under exclusive license to Springer Nature Switzerland AG 2022
J. E. Meyer, *The Post-Pandemic World*, https://doi.org/10.1007/978-3-030-91782-1_3

Deaths:
6,243,426

Fig. 3.1 Official deaths almost four million by July 2021. Including excess mortality estimates, Covid-19 deaths likely topped 10 million by mid-2021

1b. If the virus hasn't arrived within their borders, close the borders down tight. Any traveler must quarantine for 14 days.
2. Take clear action to restrict travel and set up programs to monitor outbreaks. Implement curfews and restrict gatherings.
3. Make masks mandatory and close all non-essential businesses.
4. Stay at home orders. People are allowed to go outside their homes only for food, exercise or medical treatment. Any exercise must be done within 500 m of their residence.
5. The leaders of countries which were proactive either used the advice of medical experts as the core of their response planning or handed over management of the problem to them completely.

Leaders who were slow to adopt comprehensive responses typically denied there even was a problem or tried to minimize it. They looked at the short-term economic cost of a strong response and delayed action. This was partially due to the involvement of non-medical and scientific people in the decision making process. They were reluctant to take the pain of effective measures upfront.

They failed to establish a clear decision making process and a coherent and unified message that would allow citizens and the medical infrastructure to coalesce into an effective pandemic defense (Fig. 3.2).

The Awards Section below cherry picks a variety of the examples, probably well known to the reader, which best illustrate the different government responses, and lists them accordingly.

Awards for Best Performance

- Sweden: decision making was largely driven by scientists at an early stage and they evaluated the threat and implemented a clear policy.

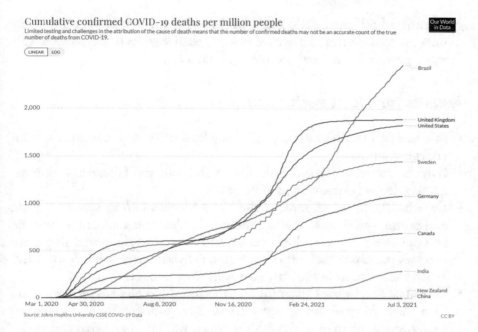

Cumulative confirmed COVID-19 deaths per million people
Limited testing and challenges in the attribution of the cause of death means that the number of confirmed deaths may not be an accurate count of the true number of deaths from COVID-19.

LINEAR LOG

Source: Johns Hopkins University CSSE COVID-19 Data   CC BY

Fig. 3.2 Relative Covid-19 deaths shows the spread of the effectiveness of government responses. Note that India death rate is now estimated to have been up to 10 times higher. (Sources: Johns Hopkins University CSSE COVID-19 Data)

- China: despite being ground zero at time zero, their rapid and decisive response (rights groups called it draconian), produced astounding results and allowed them to get back to a normal life while the rest of the world – despite having much more notice – struggled through many months of repeated shutdowns
- New Zealand – Very early response. Travel bans, lockdown, and case-tracking were effective and life returned to normal quickly.
- Taiwan: like New Zealand, world-leading performance
- Korea: Very strong and effective early response to an outbreak but later allowed a flare-up
- Australia: Travel bans, traveler quarantine, state travel bans, lockdowns with a world class result but some later flare-ups.
- Canadian Maritime Provinces: Domestic and international travel was banned. Effective tracking enabled a swift return to near-normal life.
- USA and Donald Trump: tremendously fast development of two very effective vaccines in a year was America's one bright spot.
- Dr. John Campbell's You Tube channel predicted the pandemic early. Similarly, Chris Martenson on Peak Prosperity asserted that the pandemic

had arrived by January 23, 2020, weeks before the official WHO declaration, no doubt saving the lives of many of their viewers in the process. This was expert, informed and responsible journalism.

Awards for Worst Performance

- Sweden: Early action was misguided and ineffective. A carefully considered effort but a miss.
- Italy: failure to stop travel from China and Iran and failure to lockdown quickly despite extreme rates of infection
- Canada: The initial response of the Chief Medical Officer was inadequate and counter-productive. Canada did not develop or manufacture a vaccine and was repeatedly very slow to close its borders. Despite purchasing more vaccines per capita than other countries, deliveries were often delayed and actual vaccinations in the early days were slow to occur.
- USA: Inaction by the Center for Disease Control resulted in precious time, "The Lost Month" being squandered. The response from the Federal Government, often at odds with the states, was uncoordinated and inefficient. Denial was the order of the day. Super-spreader events proliferated, and all hopes were focused on deliverance from the pandemic by a yet-undeveloped vaccine.

  - It is hard to beat the Harley Davidson Rally (2020 *AND* 2021) in Sturgis, South Dakota as an example of failed political leadership. The events featured hundreds of thousands of people arriving from all over the United States, meeting in close quarters, bars etc. without masks. They then dispersed throughout the country taking breaks every 150 km or so for food and rest stops. Would it be possible to design a more perfect method of disease spread?
  - Sturgis will have every terrorist organization in the world slapping themselves on the side of the head exclaiming "Why didn't we think of that!"

- UK: Prime Minister Boris "Can I shake your hand" Johnson almost kills himself and his pregnant fiancé with reckless personal behavior. He failed to stop international travel or to lockdown in a timely manner while Covid-19 settled in.
- Refusal of UK tourists to maintain a quarantine in Switzerland – a national disgrace – where are the jail terms, fines and losses of position and privilege?
- Spain: delayed response killed thousands of its citizens.

- Leaders from many countries who broke their own rules, some were actually punished for it.
- China: delayed warning to the world on the nature of the virus and its transmissibility.
- WHO (World Health Organization) dithered in its response and failed to issue strong warnings including labeling the pandemic as a pandemic, their most basic duty.
- Church leaders who decide that it is more important for their followers to heed their self-serving interpretation of the word of God rather than the biophysical laws that God actually laid down.
- Donald Trump whose management style of chaos was absolutely perfectly wrong for dealing with a biophysical threat such as the pandemic.

Awards for Most improved

- UK: leader Boris Johnson gets the message after his near-death experience and implements lockdowns and the affordable Oxford vaccine comes in ahead of schedule.
- Italy: very clear and decisive response but falls off the wagon months later
- Ontario's Premier Doug Ford, despite his anti-science stance and his dismantling of the province's Green Energy strategy, responded early and directly to the biophysical pandemic threat. He addressed the problem directly by taking personal charge and abiding by and promoting the best scientific advice.
- China didn't set the record for transparency but it did lead other nations in the decisiveness of its response in monitoring the public and in the rapid construction of hospital capacity.
- Canada starts late but comes back strong with an effective vaccination program

Whatever the philosophical leanings of various politicians the world over, they are now in a much more sober and down to earth frame of mind. The Covid-19 pandemic has removed a good part of the smug smile from the face of a technology and growth obsessed human society. It has left many leaders openly questioning the most basic tenets of globalism as they stood before their citizens with marginal stocks and no control whatsoever of the flow or supply of critical medical equipment from masks to gloves to medicines.

One wonders if even the Shoot-From-the-Hip President Trump would do things the same way if he had it to do over again. He might be loath to admit

Fig. 3.3 Nurses seeking covid skeptic volunteers

it, but had he embraced science on this challenge, he would likely have been re-elected. Trump didn't lose the election because of the pandemic; he lost because of his response to the pandemic (Fig. 3.3).

Back to the Beginning in Detail: A Short Summary of a Crisis, Initially Mishandled

Origins and Denial

A look at how the first days of the pandemic played out illustrates critical, but typical, failures of human responses to a large problem. Chinese medical experts and scientists were trying to get their facts straight and learn more while government officials resisted passing bad news up the chain of command and into the public domain. They tried to suppress public notifications and discussion coming coming down hard on those medical experts who felt the public should be warned.

It must be noted though that the WHO investigation team discovered the presence of many variants of the disease and reported in February 2021 that there was a much wider dispersion of cases than previously believed. This leads to the conclusion that the virus was present and spreading well before its official first case was diagnosed on December 1, 2019.

Those first few days in Wuhan, China where the Covid-19 virus first took hold and then launched itself across the globe to become the largest pandemic in over 100 years and whose death toll, at this writing, likely approaches ten million people. The excellent BBC report[1] of January 26, 2021 goes into a great deal more detail but a summary of Covid-19 s progression is as follows:

- December 1, 2019 - a 70 year old man became the first diagnosed case of coronavirus disease, Covid-19.
- December end, 2019 - several other patients appeared with the same symptoms and tests showed they were infected with a novel coronavirus similar to Sars-1 (Severe Acute Respiratory Syndrome).
- Local health and China's CDC (Center for Disease Control) were communicating on this development.
- It is possible several thousand people were infected or had had the virus but either the symptoms were too mild to report or their cases were classified as flu or pneumonia.
- December 30, 2019 at 4:00 PM the head of the emergency department at Wuhan Central Hospital received a lab report which announced that they were dealing with a Sars-1 Coronavirus. She passed this report on to her colleagues and it was leaked to the internet.
- The Wuhan Health Commission demanded that all reports of the new virus be routed to them and that leaks had to stop.
- It took a mere 12 min before these orders appeared online.
- Three hours later an editor of ProMed-mail (an organization which covers disease outbreaks) sent out an alert email on this highly infectious disease to its 80,000 global subscribers.
- One December 31, offers of help began to come into the Chinese CDC but that body gave assurances that the virus was not highly transmissible.

---

[1] https://www.bbc.com/news/world-55756452

- Mid-January 2020 – China finally releases the genetic sequences to the world scientific community.
- China's National Health Commission takes two additional days to provide a vague response to a WHO inquiry.
- China's Television refers to the doctors who leaked as "rumour mongers" and they were questioned by the Wuhan Security Bureau.
- Li Wenliang, the doctor whose warning had gone viral signed a confession and subsequently died of Covid-19 in February.

In the old Soviet Union, and occasionally possibly still in Russia, whistleblowers had a habit of hurling themselves out of the windows of tall buildings. The Chinese approach is a neater way of handling loose and noisy ends but it is a practice which does not lead to good public outcomes.

The end effect is that doctors were prevented from announcing that the new virus was a human-to-human capable disease. If it had been known that Covid-19 had infected thousands of people, this would have been self-evident but the ability of this disease to fly under the radar and infect while its hosts were asymptomatic kept it all but invisible to our detection and alert systems. Chinese officials maintained there was no evidence of human transmission well into the middle of January and they would not release the genome sequence formally until January 20.

Laboratories in China which had determined that the virus was a member of the Sars-1 family, and therefore likely transmissible, were forbidden to release their results as well as the genomic sequence. When, on January 11, Professor Zhang Yongzhen released it anyway, his lab was shut down the next day.

But after January 20, everything was out in the open. As Wang Linfa, a bat virologist at Duke-Nus Medical School in Singapore was quoted by the BBC: "January 20th is the dividing line, before that the Chinese could have done much better. After that, the rest of the world should be really on high alert and do much better." (2021, BBC).

By this time China began to launch an all-out campaign to control the virus but the rest of the world, with several notable exceptions, idly watched and dithered. By February 6, China had already built two new hospitals dedicated to the treatment of Covid-19 patients. When did world leaders begin to wonder what all the fuss was about?

Despite leading the pack on Covid response, Australia and New Zealand were slow to ramp up vaccinations. Subsequently, this was exploited by the Delta variant. Covid-19 will continue to take a heavy toll on both hospital resources and the economy until vaccines and exposure push immunity levels above the 90% threshold.

Covid-19 is now endemic in our world as the opportunity to eradicate it has been lost. Booster shots will become the norm and hopefully these will deal adequately with the series of new variants which may now be inevitable. In fact, the vaccination program is now seen by experts, not as a way of eradicating the disease, but rather as the means of greatly reducing severe sickness and death. That is because it now appears Delta or the even more infectious variant Omicron, will ultimately infect every human on the planet.

If our response had been better, how many people would have died? How many people and companies would have gone bankrupt? Perfection is not achievable but if everyone had roughly fulfilled their clearly defined responsibilities, the damage would have been reduced. If our societies were designed to absorb the pain and dislocation of groups impacted by Black Swan events, then we would be much less hesitant to take actions in the long term interest of the majority.

If Only …

But if the society is confronted by a choice between causing severe short-term pain for the certain industries and the deaths of tens of thousands of citizens and dithers, surely it is time for structural changes in our social policies. Agility matters and preparation and flexibility improve agility. But agility requires a transparent national conversation which can readily absorb new facts from experts unafraid to speak openly. In many countries, where politics and influence trump science across a number of issues, whistleblowing can amount to careericide.

Managers mostly manage steady state businesses and new challenges are outside of their job description. They don't want to report problems to their superiors and their first impulse is to figuratively declare "We've got this." Declaring an emergency may well come uncomfortably close to whistleblowing.

The table below supports the theorem that early and decisive action is the best policy in the long run regardless of the higher initial costs. The ability to make these decisions relies heavily on being able to define the threat (Table 3.1).

Comparable data for China, the origin of the virus and the country who arguably acted with the most stringent lockdowns, is not available but the graph below indicates an extremely successful lockdown. Few in the west

Table 3.1 Impacts of different Covid strategies

Canada Covid strategies			
Table: Health, liberty and prosperity impact of COVID in Taiwan, Australia and Canada			
	Health COVID cases per million (7-day average, December 21, 2020)	Prosperity 2020 GDP growth (EIU projections) (%)	Liberty (Oxford stringency metric: most constrained = 100
Taiwan	0.2	2.40	19
Australia	1	-4.10	47
Canada	176	-5.80	64

Sources: Ourworldindata.org; Economist intelligence Data

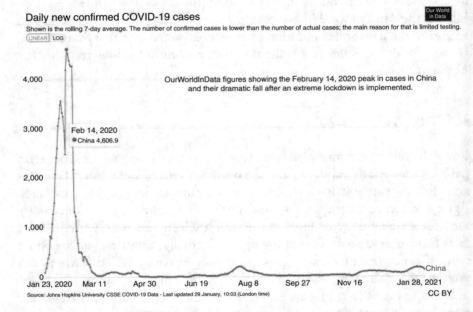

Daily new confirmed COVID-19 cases

Shown is the rolling 7-day average. The number of confirmed cases is lower than the number of actual cases; the main reason for that is limited testing.

LINEAR LOG

OurWorldInData figures showing the February 14, 2020 peak in cases in China and their dramatic fall after an extreme lockdown is implemented.

Feb 14, 2020
China 4,606.9

Source: Johns Hopkins University CSSE COVID-19 Data - Last updated 29 January, 10:03 (London time)

CC BY

Fig. 3.4 Covid-19 extreme clampdown effect. (Sources: Johns Hopkins University CSSE COVID-19 Data - Last updated 29 January 10.03 (London time))

believed the initial Chinese data and certainly China suppressed data initially, but their lockdown does appear to have been exceptionally effective.

Those countries and regions who posted the best pandemic response performance suffered much lower long term economic costs and their residents resumed close to normal life much earlier (Fig. 3.4).

Breaking Down Threats

Can the countries which did not do well against Covid-19 begin to improve their responses to sidestep the worst impacts of the next pandemic or bio-physical threat? Clearly there was no lobby promoting propagation of the pandemic but there were clear interests, large and small, directly impacted by actions taken against the virus. Maintaining social cohesion while dealing successfully with a threat requires any action taken must be fully explained. The harmful effects of that action need to be at least partially absorbed by the nation as a whole.

Resilience, self-sufficiency and good data are the best defence against being gobsmacked by a novel problem. Preparedness is built, in not added on.

"There is trade for convenience and then there is trade that grows into structural dependency. Resilience requires dependence on outside sources to be minimized." Eric Reguly, Globe and Mail European bureau chief.

Levels of Social Policy Clarity on Threats (Table 3.2)

- Immediacy – how quickly must reaction occur to avoid catastrophe. Immediate = 10 and a century = 0
- Threat Level – how much of an existential threat it constitutes

  - Titanic – 10 to passengers, no action means death in 2 h
  - Pandemic – high to affected populations, 3 to global population

- Time frame – time before a non-effective response produces catastrophic results.

Table 3.2 Comparative threats and responses

Crisis	Immediacy	Threat level	Time to disaster	Clarity of issue (%)	Reaction strength	Interest group resistance
Titanic	10	10	2 h	100	10	0
Pandemic	8	3	6 weeks	80	7	2
Climate change	4	10	30 years	75	3	7
Resource depletion	2	10	50 years	20	1	9

- Clarity of Issue – how clearly has society come to see the crisis.
- Reaction – how strong and decisive the reaction was.
- Interest groups – are there powerful interest groups who oppose decisive action.

The passengers and crew of the Titanic had minutes to respond as the ship began to sink after striking an iceberg. No one denied that the ship had hit the iceberg or was in mortal danger and there were no interest groups who would profit from the delay of emergency measures. Since the ship's designer was on board, they knew with certainty that the Titanic would be resting on the floor of the North Atlantic within 2 h. With that inevitability clear in their minds, passengers and crew then made a coordinated attempt to save as many lives as possible.

Had they delayed their response, more lives would have been lost and if the ship had been better prepared, with adequate numbers of lifeboats, fewer lives would have been lost. Had the owners of the Titanic been focused on the well-being of their passengers rather than on the commercial and status gains that would flow from setting speed records for an Atlantic crossing, it is likely the accident would never have taken place.

Why Has Covid-19 Been So Successful?

The human enterprise has become a dream environment for aspiring viruses. Humans and our animals constitute the richest target in the history of the earth for viruses. This unprecedented amount of biomass has been made possible by the use of fossil fuels that boosted food output and the exploitation of resources. Given the myriad of vectors in our society and the speed with which connections can occur, a virus can spread easily. If the virus becomes infectious before symptoms appear, it can rapidly propagate around the world before being detected.

Covid-19 filled those requirements.

What We Can Learn from the Covid-19 Pandemic

Looking at successful approaches which minimized economic damage as well as the loss of lives, we can summarize the winning approach as follows:

- Learn to identify a problem and how to interpret an exponential curve
- Blow the alarm early, activate emergency systems ASAP
- Turn the issue over to experts as soon as it is identified
- Invest heavily in acquiring the best data
- Formulate a plan for getting ahead of the problem.
- Execute it with resolve
- Examine how effective the response was and consider how it could be better.
- Don't let up until it is clear the threat has been finally defeated.

Imagine if a more transmissible variant of Covid-19 had struck China and instead of 100,000 infections, 100 million Chinese were made ill. Would China have exported medical supplies of any kind under those circumstances? Where would countries without medical supply manufacturing capacity get what they needed or would they simply do without?

Apply this to the post carbon era; where are major energy importers expecting to get their energy, their food and their manufactured goods? Climate change and declining fossil fuel reserves will deliver the kind of temporary disruptions we experienced during the 2020/22 pandemic on a much larger scale and many of them will be permanent (Fig. 3.5).

Covid-19 spread was depicted using exponential curves of its growth rate. Above is what an exponential loss curve looks like.

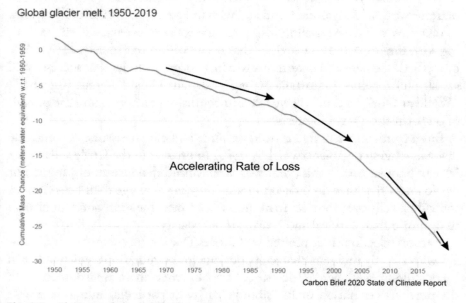

Fig. 3.5   Glacier melt is rapidly increasing

How to Fail in a Crisis

- Don't double check when the first signs of trouble occur.
- Don't alert the proper authorities.
- Hope it will blow over and do nothing.
- Put non-experts in charge, or worse, image conscious politicians
- Allow multiple sets of books and information flows
- Do not communicate from one central platform
- Do minimal testing
- Don't follow up with the potentially infected
- Don't bother determining your long term need for essential supplies.
- Don't overbuild or maintain capacity needed to cover extreme peaks in demand.

Complexity, Dynamics and Threats Which Learn

The biophysical threats of climate change and resource scarcity (foremost among them energy) are complex and, as they unfold, new challenges will arise as the knock-on effects manifest themselves. These problems will trigger an unpredictable series of events occurring over an unpredictable timespan.

We live in a dynamic world which has many systems ebbing and flowing in patterns we don't fully understand and with linkages we can't currently absorb.

Not only will we be dealing with a natural series of events, we will be dealing with the impacts and reactions that occur within our society. How stable our societies remain will determine, to a huge degree, our level of success. Will we all pull together ("Australia Strong", "Britain United", "Italy Together", "World as One", etc.) or will we fall into confusion, bickering and the pursuit of our own short term interests?

But a pandemic is different from resource and climate issues. A pandemic involves a living organism, which is trying to prosper in the fertile soils of the human population. To this point we haven't found that viruses can think, but we do know that they do indeed mutate. They can't change their behaviour as individual cells over their lifespan but, as a species, they can and will change to optimize their survival and propagation capacity.

Natural selection, first posited by Charles Darwin, occurs when individuals, with slightly different characteristics due to the mutations which occur in every child of every parent, do either better or worse in an environment than the previous generation or its "siblings". If their particular mutation allows them to better propagate, their offspring carry forward this mutation and proceed to "go forth and multiply". And when the reproduction cycle occurs in between 6 and 72 h, new mutations can get a foothold very quickly.

In this sense, viruses, as a species, can learn. This enables them to only be more effective in infecting us but also allows them to partially counter any measures, like lockdowns or vaccines which we implement. It became apparent nine months into the pandemic, variants of Covid-19 had developed greater transmissibility which allowed them to prosper in an environment of lockdowns that had been gradually defeating the original form of the virus.

The capacity of the bug to "learn" (change and improve) makes humanity's job of eradicating Covid-19 very much harder.

Fortunately, the much larger problems of climate change and resource scarcity aren't actively changing themselves to better defeat human defences. But the difficulties they present are large and complex enough to require our full and undivided attention. Now that we've effectively let the pandemic out of the box, killing it (or learning to live with it) has required several years of high levels of effort and behaviour change. But given consistent effort, post-pandemic life can largely return to a somewhat familiar normal.

Conversely, climate change and the transition to renewable energy and sustainable lifestyles will require that the changes we make will be permanent. Furthermore, the effort we make to arrive at that changed state will have to be maintained for decades.

We can't arrange to have a mini-climate change or a mini-fossil fuel collapse so Covid-19 is the next best thing to act as a rehearsal that will allow us to improve our preparedness. Covid-19 can effectively learn. The question is, can we?

When Promoting Business-as-Usual Is Counterproductive

When a person becomes infected, the body springs into action and boosts its temperature to unleash a host of virus and bacteria fighting measures. We call this "fever" and it is a natural part of healing. Although it might make us feel sick, it is actually the body trying to make us healthy.

But this effort costs energy. To raise its temperature by 1°, the body must increase its metabolic rate by 10–12.5%. That is why a person can lose several kilograms of weight over a 3 or 4 day sickness. The body is burning up fat, carbohydrates and protein, to fight off the invading threat. We may not have an appetite because digesting food takes energy and the body is busy fighting an all-hands-on-deck battle with the virus.

Fevers aren't pleasant but trying to restore normalcy by taking medicine to reduce the fever makes long term outcomes slightly worse. Bodies boost their temperatures for good reason and when the battle is won, the fever "breaks" and the return to normalcy can begin. The fever subsides and appetite returns.

Human bodies, human social systems and the planets natural systems all require healing from time to time and this process, like all maintenance, requires energy, resources and a reduction in normal performance. Hoping for a return to normal is natural but we shouldn't try to maintain normal business patterns when events are far from normal. It will result in a poorer long term outcome.

Conflict of Interests

Scientists make mistakes but science always learns from fixing them. Science is accumulated knowledge that is built first on theory, then observation and proof and fine-tuned by experience. An open national conversation promotes learning by accumulating more and better data and building transparent reporting. Policy makers and society in general have to trust science to be their guide on technical questions.

But some interests don't want questions raised or current priorities challenged. Overall well-being issues occasionally interfere and conflict with social norms and economic priorities.

In many regions of the world, people eat at night what they've earned during the day. This pattern of almost literally hand-to-mouth existence makes lockdowns difficult and is a huge drag on resiliency. Storage of even a few days' food and energy is critical for stability. If people panic because they think there might be a shortage of toilet paper, what would they do if they thought there would be a real shortage of food?

Overcapacity is critical to handle unforeseen peaks in demand for food, energy, manufacturing, hospital beds, etc. The globalism mantra of efficiency, "just-in-time" inventory needs to be replaced with localized "fast-reaction" capacity.

In all areas of the world there are powerful interests heavily invested in the continuance of business-as-usual. Resistance of these interests can be expected but it should not be allowed to derail efforts to protect the common good.

People Are Resistant to Change but People Can Change

In 2018, Doug Ford came into office as Ontario's Premier with no strategies or background studies on green energy or the hydro rates he so roundly denounced. He was intent on both terminating renewable energy initiatives and paving over the Green Belt of farmland surrounding Toronto with houses to accommodate an ever increasing level of immigration. He even tore some EV charging stations out of transit station parking lots.

His party's criticism of expensive electricity costs compared to those in the United States was so poorly researched it didn't even allow for exchange rates and thus made US costs appear 30% cheaper. Not only did he ignore grade 3 math, he sneered at science at every opportunity.

In his mind, and that of his party, food security, climate change and the inevitable transition to renewable energy were theories which might possibly inflict serious damage to the province somewhere in the distant future. In any case, the market would be able to adjust in time. Government action was neither wanted nor required. He didn't understand these biophysical issues and it was easier and far more politically expedient to ignore them.

BUT, that same Doug Ford, when confronted with images of overflowing hospital corridors and body bag shortages in Italy, suddenly was able to comprehend a biophysical threat that was going to arrive on his doorstep in days. He very quickly pivoted and began to take some very sophisticated modelling-based advice from experts and scientists. He even stopped touting endless growth as a cure-all and actually threw it under the bus in an effort to thwart the spread of Covid-19.

Was he successful? Partially. But the key point is that he absorbed scientific advice and implemented measures that were aimed at the well-being of the provinces citizens. Why can't he apply the same formula to our other greater threats?

Answer: powerful and heavily embedded interests. Ford's pandemic response was a 180° change from his previous approach of getting rid of government to allow corporations to have free reign because in his mind, free markets work better than government: except maybe for fighting a pandemic.

Will he revert to form after the pandemic subsides? It's an open question, but certainly his calls for the establishment of local manufacturing of critical materials are unlikely to be abandoned and these fly in the face of the globalism agenda he previously supported whole-heartedly.

Since he understood the threat of exponential growth of a virus, can he be that far away from connecting the social and environmental decline of his province with unsustainable population growth?

Perhaps Covid-19 was his red pill.

Making the Policy Apparatus Science Absorbent

The key to good policy is good data and full context models in the hands of scientists fully supported by open-minded politicians. Ford has the same level of data available from similarly accomplished experts in the fields of renewable energy and demographics if he chooses to use it. But, Covid-19 proved that he is capable of seeking and following expert advice in the national interest.

This is the winning formula whether the challenge is warding off a pandemic, improving equality levels or transitioning to renewable energy.

There are thousands of Doug Fords embedded in the political, media and policy-making world. They serve the status quo and the current elites. They may also genuinely believe that public welfare and the narrow interests they serve are one and the same. They have no means of evaluating success or failure other than by the effect on the bottom line benefit to the elites – typically GDP size and rate of growth.

But, as the pandemic demonstrated, these hard-liners can change. Doug Ford changed when he saw the Italian disaster unfolding hourly. He suddenly embraced the kind of scientific advice he had flatly rejected on climate change, urban sprawl and other biophysical issues.

For the UK's Boris Johnson, change required a near-death experience: his own. But he did get the message and began to implement pandemic measures in earnest.

Making biophysical crises real for the political class is a vital first step towards directly facing the challenges on our doorstep.

What We Should Do for Covid-19 or Any Crisis Once We Are in It

When we find ourselves having put off necessary action, like the proverbial frog in warming water, how do we begin to get ahead of the crisis?

1. Admit there is a crisis and that we didn't respond well initially.
2. Identify the causes.
3. Resolve to do what is necessary.

The COVID Strategic Choices Group, in their December 30, 2020 report, "A New Strategy to Protect Canadians from COVID", suggested that to make progress in the middle of the pandemic, it would be necessary to do what we should have at its beginning. Only now, the necessary steps will be more difficult and more expensive.

A Template for Success

1. A Cohesive society. People have to believe that their government has their interest at heart. Unfortunately trust in governments in the western democracies is declining and the Covid-19 performance of few of these nations is likely to reverse that trend.
2. Crush or manage? Is this a problem that can be managed or does it have to be brought to ground at the outset? A managed problem only stays managed while the problem is stable. Management requires an abundance of monitoring. Attempts at management will bring surprises in terms of a counter-punching virus.

   – Ride the dragon?
   – Viruses change and so do peoples' behavior
   – Buildings don't wear out but staff does, particularly if they are being worked hard and exposed to the danger they are treating.
   – The result is that demand for health service goes up while the number of staff available (not sick or in quarantine) goes down.
   – This is a structural shortfall that defies simple and superficial solutions.
   – If a hospital system is at 90% capacity, managing the last 10% is fraught with uncertainty unless there is a significant amount of additional capacity that can be brought on line quickly
   – The longer a virus is out there circulating the more variants it will produce and hence the greater the chance that a vaccine-defeating variant will arise.
   – The Trump approach ("What dragon?") of attempting to maintain business-as-usual and pin the nation's hopes on a vaccine (technological cure)

- Huge damage done during the waiting period of a year to 18 months
- Problem gets out of hand in the waiting period
- Problem changes in the waiting period
- Key sectors decline and weaken in the waiting period

Outside of citizens of China, New Zealand, Taiwan and Australia, most people would like their governments to do a vastly better job in dealing with the next pandemic.

But let's examine our own personal responsibility. Pandemics are a well-known threat and their more regional relatives, epidemics, happen fairly often. How well prepared were you and your family? It seems few of us were any better prepared than our governments. Are we all determined to be better prepared next time? Because there will be a next time.

Ask yourself "What can I change to make myself and my family less susceptible to the effects of the next pandemic?"

What did the people who suffered the lowest Covid-19 damage do to merit such outcomes? It comes down to lifestyle and diet. Where they live, how they live and work and what they eat and their use of drugs, tobacco and alcohol had a large impact on the severity of their Covid-19 infection.

Fit and well-nourished people who habitually looked after themselves were far more likely to have milder symptoms and stay off ventilators than people who beat themselves up by smoking and drinking while they depressed the sofa springs eating junk food.

If you want to stay healthy, breathe a lot of fresh air and eat food, mostly plants. Live and work in low density environments and stay out of mass transit including its most rapid and constrained variety, airplanes.

That is very easy to say but few can check off all of those boxes and those who can are probably well-to-do or work on the land. But, the flight of people out of dense urban cores all over the world during this pandemic left little doubt as to which way those seeking to remain healthy would jump had they the resources to do so.

The population segments which were unable to do this are the exact segments whose Covid-19 outcomes were the worst. Poor people, in densely packed housing and close quarter, low wage jobs and (sadly for public transit advocates, of which the author is one) who used mass transit as a daily part of their lives. These people were far more likely to become sick and end up on ventilators or die.

We know what does and doesn't work for avoiding plague infection and optimizing outcomes. Are we able to incorporate that learning in our own lives? Can we affect changes in government policy to assure both ourselves and our neighbours better outcomes in the next unscheduled (but certainly on nature's calendar) event?

Biophysical Threats on a Different Level

Pandemics happen from time to time but despite individual suffering and perhaps degrees of societal collapse, humanity struggled to its feet afterwards and continued up the ladder of learning. The same level of recovery can't necessarily be expected if we descend into the teeth of climate change. This is particularly true if we are as unprepared for the renewable energy transition and resource depletion as we were for Covid-19.

Regional renewable resource depletion has been a common event over the past 12,000 years. People moved or died but eventually the resource recovered and people moved back, possibly to do it all over again. Climate wiggles happen all of the time and these may be strong enough to largely remove human populations from significant areas for tens or hundreds of years but, over the past 12,000 years at least, the weather has reverted to more beneficent patterns.

Until 400 years ago, humanity had never developed or lived off a source of energy that was not embedded in its food or natural energy flows. But the energy we are dependent on now for our food, housing, goods and transport is mostly mined out of the ground from non-renewable reserves which nature will take hundreds of millions of years to replenish. Humanity has never before had to deal with the permanent removal of 80% of its most essential resource.

In addition, simultaneously occurring with the most radical shift in climate in our societal memory, resource depletion adds up to a double whammy. The combined challenge of these two threats should take the small pandemic-induced paradigm shift now occurring in the minds of many people, and push it all the way over to a fundamental rethink of the way we expect to live on this planet.

Making Societally Transformational Decisions Based on Less than Complete Information

Leaders rarely have access to the issue-specific level of scientific information they possessed when they began to make their decisions on Covid-19. The medical threat was clear very early and there was a great depth of expert advice leading them through the process because we had done this many times before. Most developed countries had commissions dedicated to infectious diseases which had been through outbreaks before and had studied pandemics intensively.

This established emergency response infrastructure had been created to deal with exactly the type of threat the Covid-19 pandemic represented. Even with that kind of expertise on-hand, the performance of most governments was, initially at least, poor. By the time leaders had cleared their heads and fully focused on the problem, in military terms, the enemy had already landed and was well dug-in.

From that point on, national health systems were playing catch-up. The public was alarmed and confused. The absolute clarity that was present in the initial days of the crisis became muddied as the need for sacrifice became larger and the length of time sacrifice needed to be made was extended repeatedly.

Upsetting the Natural Order

The Covid-19 pandemic upset the natural order our societies had become used to. This natural order was based entirely on human priorities. All of a sudden something came along that showed twenty-first century humanity that we cannot assume that all outcomes will be in our favour.

Humanity is in a relationship with nature and the foundation of every relationship is how well the partners know themselves and how well they understand the other's needs. The pandemic revealed that we didn't understand ourselves very well and that the socio-economic system we've built up in the last century was very poorly suited to dealing with a significant biophysical threat.

We are learning to cope with the coronavirus and have made many alterations to the way we live and do business. Some of these changes will be permanent. Some of the changes have been extremely inconvenient and will be discontinued as soon as it is safe to do so. But others have turned out to be simply better. Space, clean air, family time, zero commutes, regular exercise, improved working conditions, low density housing and healthy eating are but a few of the beneficial changes Covid-19 embattled populations have experienced.

People will want to hold on to these. They will also want to change the things in the failed globalized economic system which effectively threatened their health and well-being.

An attachment to nature reappeared on the radar for many people although possibly for more of an escape than as a vital partner.

But nature is indeed our vital partner and we need her much more than she needs us. As the pandemic battle winds down, we need to absorb the magnitude of the task of transitioning to renewable energy and contemplate the ramifications of living as a sustainable society. We will need to become more aware of ourselves and of the needs and workings of our very significant other, Mother Nature.

4

Our Conversation with Mother Nature

Ultimately, all life forms will end up living on the flows of energy through natures myriad systems. Some may be able to exploit stocks accumulated over years or eons such as fossil fuels, but these are finite reserves and will deplete as they are exploited. Forests, fisheries and soils are renewable systems and, with prudent management, can be maintained indefinitely.

Once depleted, forests (depending on soil health) can regrow to their former glory in time spans of 100–1000 years, fisheries can rebound in decades and some agricultural soils in just a matter of years under ideal conditions. Minerals and fossil fuels can only be taken out of the ground once. Minerals can only be recovered indirectly by recycling and combusted fossil fuels won't be replenished for tens of millions of years.

Accordingly, once we have past the blip of fossil fuel use, mined the richest mineral deposits, destroyed much of the biological systems which have enabled us to leap from fire apes to interplanetary explorers in a mere 12,000 years, we have to ask ourselves; "What comes next?"

The rich resources we consumed getting here are now substantially depleted and the resource pickings will be slimmer. Squeezing blood (or minerals) from a very low ore grade stone takes a lot more energy and produces a lot more waste. Although we have become a great deal more accomplished technically, we have innovated ourselves into a bit of a corner that only a strategic change will get us out of.

The countries who did the best against the pandemic used tactics from the Middle Ages; they shut the gates and identified and isolated the infected. Technology nibbled around the edges in the form of improved treatments but failed completely to deliver a knockout blow until at least 2 years into the

© The Author(s), under exclusive license to Springer Nature Switzerland AG 2022
J. E. Meyer, *The Post-Pandemic World*, https://doi.org/10.1007/978-3-030-91782-1_4

event when mass vaccine injections were performed. Countries which either attempted to manage infection levels or who stood back and waited for the what they hoped was the decisive medical option, vaccines, simply failed to protect their citizens' lives and financial security.

Tried and true common sense approaches weren't acceptable to the business-as-usual leaders and, in the end, they unintentionally delivered the worst of all worlds in terms of economic and human losses.

The larger issues of climate change and resource depletion are no more amenable to technological fixes than was the pandemic. Technology can definitely be a huge help but if the right structural decisions are not made, they can only delay the inevitable.

We have to realize that from this point forward we will be increasingly living off not what we can pull out of Nature's cupboard, but what we can take off her fields. This will seem like a hand-to-mouth existence but flows are ultimately what we have to live off and we need to understand them thoroughly.

A Failure to Communicate

Mother Nature speaks and listens in physical units. Prayers, political promises and GDP growth reports don't register on nature's radar.

Conversely, in 2021, most human feedback loops take place in the human domain driven by energy we produce ourselves by digging it out of the ground. Nature is almost invisible in this environment and completely disappears in the money economy.

For the past 400 years, set free by fossil fuels, human society has done things on the basis of "because we can" but will now have to begin to operate along the "because we should" theme.

If we are to deal with nature, we have to learn more about ourselves. Self-examination indicates that a certain percentage of people will not comply with any new standards of conduct unless they are forced to. They'll adhere to social norms once they are established but will oppose any change. Civil strife will send change strategies off the rails. Consequently, social friction can only be avoided by minimizing inequality and by involving all segments in society in a fact-based conversation.

People are both the problem and the solution and everyone needs to take stock of their own situation to understand how they can best avoid disruption. Getting ahead of the transition to renewable energy and building a healthy, resilient way of living requires that a fundamental re-defining of "normal" is in order.

And, as people move their own lives toward the path of sustainability, they must, at the same time, be making efforts to see that their community and nation are starting to move in the same direction.

Getting to Know Our Planet

As we enter the third decade of the twenty-first century, our science has built up an extensive body of information allowing us to grasp the magnitude of the standing stocks present in earth's biosphere. This snapshot of the current situation does not mean we can understand the true health or interdependencies of the current system but it does represent an evolving level of monitoring capability.

One of the earliest and best studies with a worldwide scope was the pioneering "Conquest of the Land Through Seven Thousand Years"[1] by W. C. Lowdermilk of the U. S. Department of Agriculture, who travelled around the world in the late 1930s cataloging the decline of agricultural soils. This remarkable effort cries out for similar illumination to be cast on the other major elements of our biosphere from fisheries to forests to wildlife. If an individual had but one hour to spend to give themselves an overview of the connection between societal collapse and environmental decline, that hour could be spent in no better way than by reading this fascinating and accessible study.

It takes a community to raise a child and it takes great swaths of healthy forests and coastlines to raise apex predators like eagles, bears and the network of fauna and flora they depend on. The same holds true for any species which require a flexible range to support themselves whether it be in the Americas, Asia, Africa or Europe. Nature needs unimpeded space to allow its "natural support net" to fully develop.

Soil, forests and fisheries are complex and mutually supportive communities of different organisms and energy flows. The song they sing individually can be heard by the current level of our science but we are seemingly deaf to the full choir.

---

[1] https://soilandhealth.org/wp-content/uploads/01aglibrary/010119lowdermilk.usda/cls.html

Taking Stock of Mother Nature

The best way to communicate with nature is to listen, to take nature's pulse. We have to deal with nature as it is now, not as it was in the past before we began consuming rich accumulated biological and mineral assets.

Taking the planet's temperature requires more than a thermometer as we know that both oceans and atmosphere are warming but, beyond that, basic energy flows are also being altered. Ocean currents and Hadley cells[2] are changing with large implications for rainfall patterns and extreme weather events.

If a snapshot[3] were taken of the world in 2021, it would show humans in control of many of the earth's biosystems. Trends are more important than snapshots and according to the a 2011 biomass census complied by Vaclav Smil[4], human activity over the last 5000 years has reduced total global biomass by about 50% (Fig. 4.1).

Biosystem stocks and flows, in the form of mammal, fishery and forest biomass are changing as are their rates of growth. Trees talk to each other. Why can't we talk to trees or at least comprehend their conversations?

Well at least some of us are trying. Ecologist Suzanne Simard[5] has discovered that through their "wood wide web" trees not only talk to each other, but collaborate for mutual benefit and protection with other plant and fungi life species. This isn't just the stuff of Hollywood movies like Avatar, which used Simard's work as inspiration; it is a science in its infancy. Ancient Greeks understood that the planet's soil assets were depleting as have every subsequent society. We need to achieve the same level of understanding with regard to all of the major human support systems, which of course, support all species.

Location, Location, Location

Where one lives has a great deal to do with the amount of energy needed to live and it also has a great deal to do with the amount of energy that is practical to generate and store.

---

[2] Hadley Cells are the low-latitude overturning circulations that have air rising at the equator and air sinking at roughly 30° latitude. They transport moisture from the tropics northward and are responsible for the trade winds. www.seas.harvard.edu/climate/eli/research/equable/hadley.html

[3] https://www.greenpeace.org/international/story/17788/how-much-of-earths-biomass-is-affected-by-humans/

[4] www.vaclavsmil.com/wp-content/uploads/PDR37-4.Smil_.pgs613-636.pdf

[5] https://www.youtube.com/watch?v=Un2yBgIAxYs

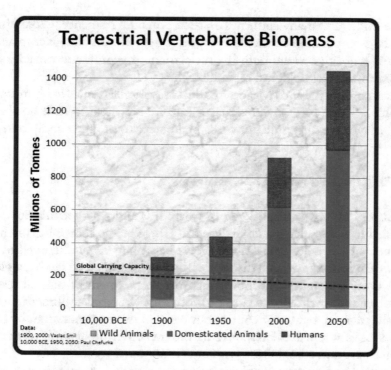

Fig. 4.1 Fossil fuels have allowed us to dominate and displace earth's natural systems

Latitude matters but so, to a lesser degree, does longitude. Residents of the west coast of Europe benefit from the warmth and moderation of the Gulf Stream – the planets biggest heat transfer mechanism. The same kind of mechanism of air and ocean currents also has a large beneficial effect on the central and northern west coast of North America.

People in the middle of the continents, are subject to greater temperature extremes accompanied by more sun and less humidity. It all counts in the energy demand/energy production balance.

New Zealand has been rated as the most resilient nation in one July, 2021 study.[6] The Global Sustainability Institute study[7] looked at the collapse of supply chains, international agreements and global financial structures. It found that a combination of ecological destruction and population growth could trigger a decline in the sophistication of our societies over decades; a very short period of time.

---

[6] https://aru.ac.uk/news/new-zealand-is-best-placed-to-survive-collapse
[7] https://www.mdpi.com/2071-1050/13/15/8161

The study rates many countries on their potential resilience. Close proximity to large population centres is a very large negative while high cropland per capita is a positive as are both hydroelectric and geothermal capacity along with advanced technological capability.

But according to Linda Mearns,[8] of the US National Centre for Atmospheric Research, "I don't see any area that is safe … Nowhere to run, nowhere to hide." Maybe we can't hide but we can take steps to properly equip ourselves.

Getting to Know Ourselves

The history of how humanity got to where it is and whether this rise can be extended in perpetuity is a large question. There are no remaining uncharted continents to be colonized nor are there likely many large untapped reserves of critical resources awaiting discovery. The low hanging fruit has been picked.

We need to comprehend how much the despoilment of natural resource stores has underwritten our current prosperity. We need to understand why we fight wars on an almost continuous basis. Is this an innate trait of Homo sapiens or is it a natural response to the depletion or overuse of our own resource base and competition with our neighbours?

Is eternal peace a matter of humans attempting to curb their innate aggressiveness and hostility or is it a matter of humans finally developing a social memory long enough to see why wars happen. Understanding that process would enable us to look forward and take the steps necessary to prevent the conditions which make war inevitable from occurring.

- What is real and what isn't:

  - It is commonly recognized that the commercial economy does not represent the health of the environment. Increasingly, it is evident to most that neither does the size or rate of GDP growth indicate social or personal well-being.
  - Beyond that, is the growth of the finance industry which has become a large part of most western economies. The term "real economy" is now being used to describe the portion of the commercial market that produces real goods and services as distinct from the finance economy

[8] https://www.euronews.com/2021/08/09/watch-live-un-scientists-release-monumental-climate-change-report

which adds little or no real value but deals in asset inflation, transfers and baseless money printing.
- Public discussion needs to differentiate among these and focus on the real wealth production process and human and environmental health.

- Built in Conflict

  - Some people profit from selling off the commons that all people need to sustain themselves. Whether it is paving farmland with subdivisions in the twenty-first century in Canada and Australia or cutting down and selling off the great forests of the Eastern Mediterranean 3000 years ago, short term individual profit has been in perpetual conflict with long term social welfare.
  - This conflict occurs as societies become stratified and interests diverge.
  - In hunter gatherer societies, members of the band had common interests and roughly equal shares of the commons. Equal shares mean equal interest and thus a greater likelihood of coherent policy.

    - In our current highly stratified society, equal shares of the commons or material production may be impossible. Perhaps equal levels of opportunity, information and common references should be the basis for public policy.
    - Ownership and responsibility to the future spawns obligation to seek a balance between human activity and natural capacity. Unhinged commercial economics holds that market forces and technological advances in the nick of time will provide solutions to any issue and guarantee growth forever.

Coherent policy requires a more holistic view. The actions of a broad spectrum of national and regional governments during the Coivd-19 pandemic demonstrated that leaders can indeed look beyond the commercial market place when a threat becomes sufficiently obvious.

Energy and Resource Demand of an Expanding Human Enterprise

Human energy and material consumption has increased along with our technology[9] and ability to exploit resources (Fig. 4.2).

An electric society will operate substantially differently from one based on fossil fuels. There are some things that electricity cannot do as well as fossil fuels but, in the end, electrical systems are vastly more efficient. Consequently, the amount of energy a society needs to sustain itself will be much lower than in the past. As we learn to become more efficient and to work with both

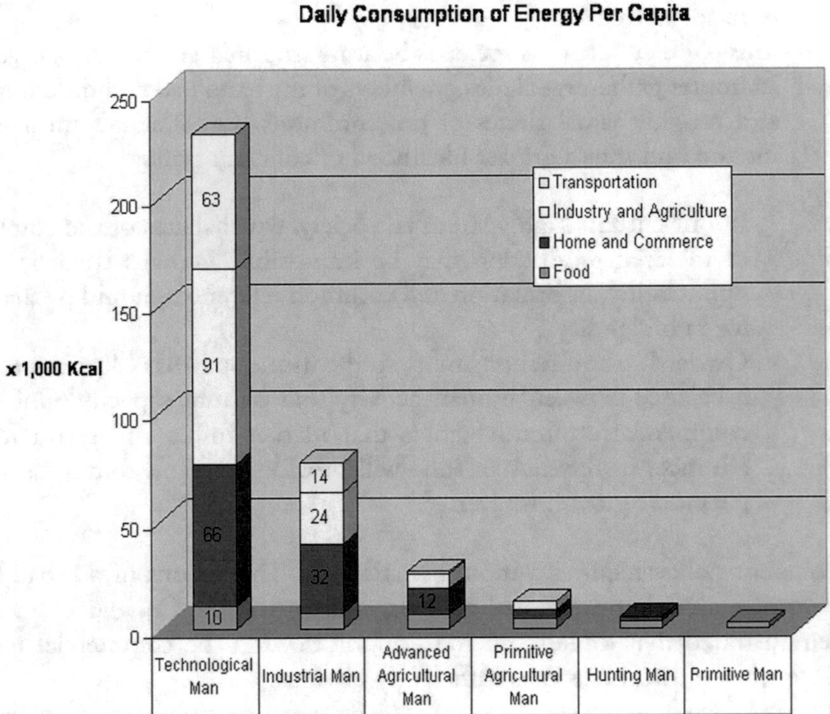

Daily Consumption of Energy Per Capita

	Technological Man	Industrial Man	Advanced Agricultural Man	Primitive Agricultural Man	Hunting Man	Primitive Man
☐ Transportation	63	14	1			
☐ Industry and Agriculture	91	24	7	4		
■ Home and Commerce	66	32	12	4	2	
▦ Food	10	7	6	4	3	2

Fig. 4.2 How we have spent our energy budgets as more energy became available

---

[9] https://people.wou.edu/~courtna/GS361/electricity%20generation/HistoricalPerspectives.htm

smaller amounts and lower grades of energy we may find that prosperity can thrive alongside the earth's natural systems. That is, if we step away from the paradigm of ever increasing population and consumption growth.

Economies of scale means that the more that is produced, the lower the unit costs will be. As human societies transformed from agrarian to industrial and energy costs kept dropping, economies of scale delivered huge increases in material wealth for falling levels of human labour.

But things have changed. Now in a world of substantially drawn down resources, we have to do more work to produce the food and material we need (or want). As our needs and population have increased, the difficulty of extracting resources has gone up and therefore costs per unit of output have increased. This is dis-economies of scale and it represents a very different reality from the one which lasted for the 300 or 400 years of the fossil fuel boom.

Commercial economic theory may indicate costs will be lower the larger the consumption but the reality of the biophysical world indicates just the opposite. Now we have exploited the richest resources, everything we mine in the future will be lower grade and planners will have to deal with the global shift from economies of scale to dis-economies of scale.

Efficiency to the Rescue?

The transition to a largely all-electric society will see energy consumption drop dramatically but there will not necessarily be a significant drop in mobility or comfort. Material consumption though must go down due to resource decline and scarcity. Easily reparable and upgradable goods with longer lifespans and high recyclability will reduce demand for resources and energy.

That same technology has also given us the ability to do more with the resources we consume and, although efficiency alone won't save the day, it will give us some level of buffer against resource crunches and the societal stress which follows.

How Have Things Been Going?

The period from 1700 to the 1950s saw a rocket-like trajectory of population and consumption growth accompanied by tremendous social and scientific progress. Over the past 6 decades however, high rates of growth haven't necessarily been beneficial for the economic welfare of the majority of people in the western world. Debt levels have increased, job quality has declined and housing has become less affordable.

Bigger isn't better for peoples' quality of life as this is illustrated by the life satisfaction levels of Canadians in the largest and fastest growing urban centres (Fig. 4.3).

Similarly, indices such as the Genuine Progress Indicator show that despite large increases in GDP, human well-being is being left behind. Canada has seen one of the fastest rates of economic growth in the OECD from 1960 onward but its level of income equality has declined from second best in the world to the mid-20s.

Below, the disconnect between the well-being of Americans and the size of the US GDP is very clear (Fig. 4.4).

But the political system is still fixated on growth based on an assumption that ceased to be true 2 generations ago. In what seems to be a typical example, a survey of residents in Orillia, a town in Ontario, found citizen priorities to be:

1. Quality of Life – 26.5
2. Healthy Environment – 24.3
3. Sustainable Growth – 19.4

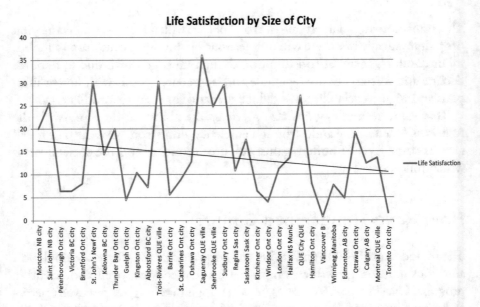

Fig. 4.3 Bigger isn't better for most people when it comes to city size or growth rates in Canada

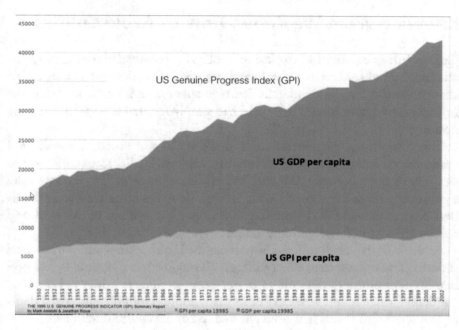

Fig. 4.4 Decades of growth in the commercial market have failed to deliver progress for most people. (Canadian Club of Rome)

4. Professional Progressive City – 16.2
5. Vibrant Waterfront – 15.5
6. Heritage Core – 14.7

Citizen preferences and priorities were all for a healthy, natural environment which are the diametric opposite of the effects population growth has. Public responses were duly noted in the survey report but the recommendations of the report then sloughed these aside and recommended ways to accommodate higher rates of growth.

The statement from a local mayor sums up the general feeling of municipal politicians, the majority of whom receive the largest proportion of their election campaign funds from developers:

"It is a closely held belief that growth is good for municipalities."

For some, more subdivisions built on prime farmland are clearly very good but, especially in the era of the transition to sustainability, continued growth is a disaster for citizens, society and the environment.

Knowing What We Know, How to Avoid Conflict

Change brings conflict and confusion so our very comfortable consumer societies need good reasons to embark on a program of fundamental change. But change we did for the pandemic. Most people are quite clear on what the impacts would have been if the world had simply ignored Covid-19 and done nothing.

The virus would have spread around the world and mutated at will until everyone was immune to all of its variants. The case fatality rate (CFR) would easily have been between 1% and 2% assuming none of the variants were significantly more lethal than Delta. This would mean 80 million to 160 million dead worldwide, 7–14 million in Europe, 3–6 million in the USA and maybe 15–30 million dead in both China and India.

Aside from the death toll, the impacts to the health system and the cost and suffering due to long term effect sufferers ("Long-Termers") would be staggering. As economically difficult as it has been to operate while fighting the pandemic, there might well have been vastly more economic and social damage inflicted if the virus had been allowed to run its course unchecked.

Don't Use More Than the Earth Can Sustain (Fig. 4.5)

How many days would it be before the earth's population would exceed the planet's sustainable output if we all consumed at the level of these countries?[10] Northern countries and heavy producers of fossil fuels go into overshoot much more quickly than countries with much more moderate climates.

If everyone consumed at the level of the average American, we exceed the annual biologically sustainable level of consumption in only 72 days. Indonesian levels of consumption are almost sustainable and take 352 days to exceed earths 365 day production capacity.

Success requires that all of the remediation programs we undertake have to be aimed at reducing our consumption levels to fit within sustainable limits.

---

[10] https://www.overshootday.org/about/

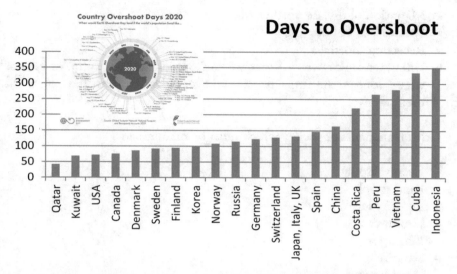

Fig. 4.5 Most countries blow through their sustainable ecological budget by mid-year

Threats on Deck

Our focus was placed firmly on the pandemic for almost 2 years but the environmental challenges we faced before it arrived have slowly worsened. The threats and challenges climate change and resource depletion present might be different but they fortunately are subject to many of the same cures.

We need to substantially reduce the impact of the human enterprise on the planet. We can do this with the twin approaches of reducing our per capita consumption levels in addition to stopping population growth. Replacing fossil fuels with electricity will make us far more efficient but the transition will be a challenge on its own.

Listed below are a few of the challenges we have to address and below that is a list of basic elements of society that must be maintained to preserve adequate social stability.

- The obvious biophysical shocks coming our way:

  - Climate change
  - End of oil
  - Resource decline
  - Food supply threats are a combination of energy and climate issues

- Migration of disease northward
- Species loss, habitat loss
- Recycling shortfalls
- Unknown unknowns - What we can't anticipate in a complex system where long term impacts can take decades or centuries to reveal themselves.

- Critical elements of society which must be maintained

  - Energy supply
  - Food supply
  - Equality levels
  - Health care
  - Education
  - Security
  - Fiscal balance

    - In just one short pandemic, we have accumulated debt that will take decades to pay down. What happens if another crisis strikes before mid-century? What fiscal reserves do we possess? When does money printing cease to be effective?

While the above lists may suggest it will be impossible to juggle that many crisis balls in the air, what is absolutely clear is that we must get an early start in order to make changes as gradually as possible.

Let's not underestimate the magnitude of the challenge of replacing fossil fuels in our society (Fig. 4.6).[11]

If the most powerful interests in our society became rich by saving resources and reversing environmental decline, achieving a sustainable balance with nature would be a great deal easier. But to date, the wealth that has been accumulated in a small number of hands has come largely through the drawing down of mineral reserves and the degradation of the health of the biosphere. In short, change is required in the way human society functions and this will produce conflicts with the established interests as well as the every-day consumer.

Compared to the pandemic, the timeframe may appear less urgent but the pathway is also far less defined. Having gone through the war with Covid-19, surely we are now very clear on the steps we should have taken. These were

---

[11] https://ourfiniteworld.com/2012/03/12/world-energy-consumption-since-1820-in-charts/

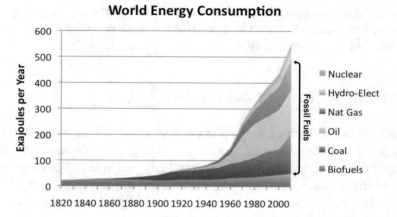

Fig. 4.6 Decarbonizing will leave a huge hole to be filled in our energy supply

very simple and very direct initiatives and basically, all governments knew what to do before the pandemic but few executed those necessary measures well.

Suffice it to say that if another pandemic threat arises in 5 years, the worldwide response will be vastly more effective and, in the misplaced words of President Donald Trump, "The virus will go away like magic." The magic, of course, will be due to an effective pandemic response.

Many societies in history have gone through resource crashes but few have seen the crash coming and managed successfully to minimize their worst impacts. Early Viking settlers in Iceland quickly became aware that the soil of Iceland reacted very differently to their traditional farming methods developed over centuries in Scandinavia. The Icelandic ground cover simply began to peel back and disappear when stressed by sheep and cattle grazing. Once begun, this process was very hard to stop, and it took concerted effort over centuries to establish a sustainable balance.

Similarly, Japan in the Middle Ages experienced very high rates of forest decline and it took a national effort to reverse it. Sweden learned this lesson late and decimated its forests by the mid-1800s in the name of iron production. Consequently, over several decades, 1 in 7 Swedes either emigrated or starved to death. A very progressive silviculture strategy was then put in place.

Hunter-gatherers would match the demand they placed on their environment by patterns of migration, along with infanticide, late marriage and low level conflicts with neighbouring tribes to keep their populations at a supportable level. The advantage of societies which are 100% focused on their immediate surroundings is that the path of sustainability is very clear.

The complex and stratified societies of the twenty-first century simply lack that build-in connection with nature. And when we do build it back into our decision making process, we'll find it involves difficult choices.

Confront Nature or Avoid Conflict

As difficult as cutting back on carbon emissions is, it will be much easier than attempting to maintain social cohesion in an aggressively dynamic climate regime. We can't fight nature on all fronts. Perhaps a business-as-usual regime can post the occasional small victory over natural threats but these will be pyrrhic in nature. Ultimately the war will be lost when nature plays her larger cards of climate change and resource collapse.

The growth dependent sectors of finance, development and speculation need that constant growth to avoid collapse and therefore are in direct conflict with the way the natural world operates. Stability and balance are not viable goals for a money and debt based system. Natural systems fluctuate within ranges and these can't be expanded without either becoming prone to collapse or taking a toll on other supporting systems.

Attempts to monetize nature simply fail to begin to address the complexity and temporal range of the biophysical world. It is much better to use biophysical economics to set limits of exploitation and allow the commercial market to operate within those limits. Fiat currency may represent the spot value of a good in the commercial market but it will never be able to represent the processes or health of natural systems currently or at any specific point of time in the future.

We need metrics which do not completely ignore the larger world around us.

The Eleventh Commandment

Lowdermilk's eleventh Commandment reads as follows:

Thou shalt inherit the holy earth as a faithful steward, conserving its resources and productivity from generation to generation. Thou shalt safeguard thy fields from soil erosion, thy living waters from drying up, thy forests from desolation, and protect thy hills from overgrazing by thy herds, that thy descendants may have abundance forever. If any shall fail in this stewardship of the land thy fruitful fields shall become sterile stony ground and wasting gullies, and thy descendants shall decrease and live in poverty or perish from off the face of the earth.

This commandment was engrained in the culture of hunter gatherers. Once humans graduated to widespread and then industrial scale agriculture with disconnected elites, this bio-consciousness faded into the milling crowd of live-for-today markets. The failure to adhere to this basic tenet wrote thousands of years of famines, wars and social upheaval into human history. When our activities conflict with those of Mother Nature, we are effectively shooting ourselves in the foot. And now, given our profligate use of fossil fuels, we've managed to upgrade our sidearm from a single shot pistol to a 30 round, fully automatic rifle with no safety mechanism. We need to de-escalate our confrontation with the natural systems upon which we are still, despite what growth promoters tell us, absolutely dependent.

A Lasting Relationship

Once we understand ourselves and our relationship with the biophysical world, we can begin to map out our options. Being able to put numbers to what were once very nebulous concepts and making sense of complex interdependencies will allow us to do and learn at the same time. We will have created a pathway for progress.

In the past, our focus was on simply getting more because, for the past 4 centuries, getting more was relatively easy. But we have developed no science, strategies or goals in regard to our place in the natural world. Consequently, as we begin the slow climb out of the Covid-19 era, we have no pathway, measurement or management systems in place to deal with the oncoming challenges.

We do, however, have a new awareness and the pandemic learning template which we can use to apply the science and social policy expertise we have accumulated to our looming existential environmental threats.

5

The Math of Our Energy-Dependent Existence

Warning! No one with less than Grade 3 math skills should attempt to read this chapter.

Numbers Spell It Out

Developing the ability to quantify an issue is a big step towards good outcomes and energy is something that lends itself well to physical metrics. Whether you are rationalizing your own energy use or submitting proposals to various levels of government or community organizations, sketching out the issue in real physical units forms a very solid basis for discussion and progress.

It also promotes good management because it makes it possible to monitor progress and modify the systems going forward. A good set of numbers allows the best decisions to be made and promotes rapid learning for everyone involved.

Our lives are basically living labs. The planet and its biosystems are dynamic and several centuries of history is no guarantee of future trends. Understanding the structure and the real-time flows of energy in human made systems and the natural biosystems around us is critical to achieving sustainable resilience.

The key questions are:

- How much energy are we using
- How efficiently are we using it
- How much energy does it take to produce the energy we are using

© The Author(s), under exclusive license to Springer Nature Switzerland AG 2022
J. E. Meyer, *The Post-Pandemic World*, https://doi.org/10.1007/978-3-030-91782-1_5

- How much damage are we doing to the planet
- What levels of consumption will the planet sustain

Knowing these things will allow the development of both a strategy and a budget process which learn over time. As we've seen with the greatly varying national results of the most recent biophysical problem, Covid-19, is that no measurement, no math, no goals = no pathway and poor outcomes.

So we'll use very basic arithmetic to be able to add scale to the issues. Instead of just asking "How much does it cost?" we'll be trying to determine how energy and biophysical systems work and how we can best work together with them.

Ratios Lend Perspective

You use ratios every day like 5% return (1:20) on an investment, and for figuring out hundreds of things from sales tax to the number of calories in a jar of peanut butter. We measure vehicle efficiency in miles per gallon, litres per 100 km or kWh/100 km.

For energy discussions there are three key ratios to understand.

- Efficiency
- Efficiency is a measure of the total energy used for a process and the total end work the process performs. For instance, a litre of gasoline contains the equivalent of about 10 kWh of energy but when burned in an internal combustion engine (ICE), produces only about 3.5 kWh of mechanical force to propel the car down the road. The other 6.5 kWh of energy is expended on internal friction and heat from the exhaust gases making it only 35% efficient.
- The efficiency of an electric heater is 100% since heat is the only output. Efficiency is important to understand because we have indeed been getting much more efficient over the past 4 centuries. We need to be able to continue to do more with less as the richness of our natural support systems declines.
- EROI or Energy Returned on Energy Invested
- EROI is a ratio of the energy put into the production of energy compared to the energy that production yields. In the glory days of virgin oil fields it was possible to get 100 barrels of oil from an energy input of only 1 barrel of oil (or the equivalent of one barrel). The EROI of solar PV is close to 8:1

and wind generation is in the 15:1 area while the oil sands set the low
bar at 3.5:1.

- This is a critical metric to understand because our current lifestyle and
future aspirations have been based on the exploitation of very rich, virgin
resources. These expectations will have to be adjusted in the light of depleting resources.
- COP or Coefficient of Performance
- COP is a ratio used to determine the output efficiency of a heat pump
system. Heat pumps transfer heat from one body of material to another. In
the case of an air-to-air heat pump, the heat pump pulls outside air through
and removes as much heat as possible and transfers it inside the house.
- If it takes 1 kW of electricity to run the heat pump and it outputs 4 kW of
heat, the ratio of energy produced to the energy consumed is 4:1. A COP
of 4:1 would equate to an efficiency level of 400%. Processes which transfer
heat can collect and transfer many times more heat energy than the energy
it takes to run the system. Making heat pumps – the transfer of energy –
more efficient is going to be critical for reducing the energy required for
heating and cooling our buildings.
- These three ratios will be used to flesh out our developmental history and
illuminate our current energy situation and possible paths forward.

Efficiency

How much of the energy used for an application actually does the job it is
supposed to do? Internal combustion engines approach 35% efficiency but
EVs are able to convert well over 90% of the electrical energy in their batteries
into mechanical work with a loss to heat of less than 10%.

For 99.9% of human existence, our only source of direct energy has been
fire. The energy potential in wood gathered for an outdoor bonfire literally
goes up in smoke as most of the useful heat convects upwards. Further, since
the fire is so unconstrained, and its temperature stays relatively low, the wood
is not completely burned and particles and embers escape up the rapidly cooling column of smoke still partially intact. In terms of cooking food or warming those huddled around the open fire, this results in efficiency levels in the
low single digit range (Fig. 5.1).

The heat from a fire in a partially closed space like a yurt, wigwam or an
Iroquois longhouse is contained much better and produces a great deal more
usable heat. Of course, also trapped inside the structure are particulate waste
and noxious gasses.

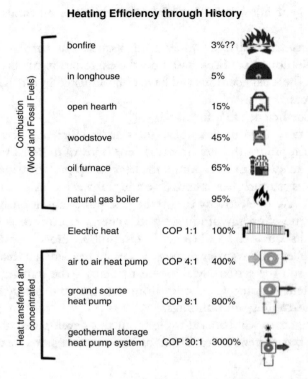

Fig. 5.1 Geothermal systems will allow huge efficiency gains to be realized in building heating

A hearth controls the airflow around the fire and allows much higher temperatures to be achieved thus combusting a higher percentage of the wood fibre. It also serves to project the heat into a closed building while its stone side both store and radiate heat.

A modern high efficiency woodstove is designed to control airflow and boost the combustion temperature so a very high proportion of the woods heat potential is extracted. Efficiency ratings of these woodstoves can approach 50%.

The carbon content of fossil fuels is much higher than wood and therefore they are able to burn much more efficiently. The combustion process results in carbon being combined with oxygen to form carbon monoxide, a chemical reaction that releases heat. CO, once exposed to free oxygen in the atmosphere, quickly becomes CO_2 or carbon dioxide.

The carbon content of wood is around 40% while coal ranges from 45% to 80%. Natural gas and propane are in the region of 80% carbon and oil can range up to 85%. Hence, when burned, fossil fuels offer a far greater proportion of combustible material than wood and therefore will be able to burn more efficiently. Additionally, fossil fuels, being gases, liquids or small lumps are much easier to precisely meter into the combustion chamber than wood for even more complete oxidation.

The New World of Electricity

Electricity is a very different energy source than those which rely on a combustion process to release energy. Electricity is the transfer of energy from one point to another and in the process the energy is drawn off as it does work.

In a combustion process, electrons are transferred between atoms as they form molecules, releasing heat whereas electricity sees the pure flow of electrons with no change in or involvement of molecules. Not only does electricity "cut out the middleman", it also permits incredibly fine control over the flow of energy.

And it also allows us to do a wider variety of work than is possible with a combustion process. In looking at the different characteristics of electrons vs fire it becomes readily apparent that the transition to a 100% electrically based economy will change the way we live and work (Fig. 5.2).

When you turn on an electric heater it achieves 100% efficiency first time, every time. With no other means of releasing its energy, an electric heater has no choice but to be 100% efficient. But that is only the start of what electricity can offer when it comes to heating systems.

The efficiencies of fossil fuels and electricity are very different. Fossil fuel offers energy anytime, anywhere (there is oxygen) in the dead of night or still of wind. But transforming that heat energy into mechanical motion is complicated as this cutaway drawing of an internal combustion engine shows.

In order to convert the heat of combustion to mechanical motion, the friction of some 2000 moving parts,[1] must be overcome and the air/fuel mixture must be compressed and then burned. This produces a great deal of heat which is lost through the exhaust and the radiator. This waste heat adds up to about 2/3 of the energy potential of the fuel.

A drawing of an electric motor is far less interesting as it has only two main bearings, one on either side of the rotating armature (Fig. 5.3).

Further differences between fire and electrons as energy sources include:

---

[1] https://medium.com/swlh/internal-combustion-engine-and-the-four-stroke-engine-12381fc54a98

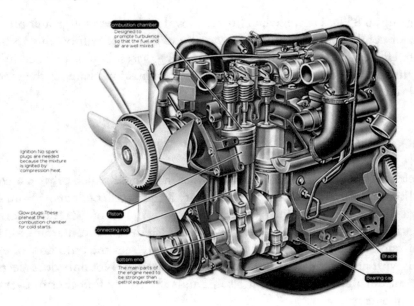

Fig. 5.2 The internal combustion engine is a miracle of technological achievement but from an efficiency point of view it is a Rube Goldberg contraption

1 moving part, 2 bearings

Fig. 5.3 Mechanically, an electric motor could not be more simple and this is borne out by its 95 + % efficiency levels

- The ability of electricity to function underwater, on land and in space.
- Electrons can power very small mechanisms while the first internal combustion shaver has yet to be made practical.

- The enormous variety of electronic machines now includes thinking, research and test machines, data acquisition, information recording, light generation and minute and focused heat production.
- Fossil fuels can only provide heat through burning while electricity can provide heat through transfer and pumping/extraction.

100% Is Only the Beginning

Clearly nothing is ever going to be more than 100% efficient because it is impossible to get more out of something than the potential energy it contained at the outset. But electricity allows us to not only produce energy; it allows us to transfer energy from one place in one material to another place in another material.

This is what refrigerators and air conditioners do by compressing a gas and circulating it. Gas gets hotter when it is compressed and cooler when it expands. The heat or cold is then drawn off and moved to where it is needed. The cold goes into the refrigerator and the heat goes into the room.

An air-source heat pump is like a refrigerator outside which takes the heat from the ambient air and pumps it inside leaving the outside air slightly colder. A ground source heat pump has its heating coils in the ground and takes heat from the earth and transfers into the house. The same mechanism can work if there is a body of water nearby.

If heat can be moved from outside to inside the building, then it is simple to reverse the process and expel heat in the hotter summer months. Heat pumps move energy from one spot to another, they do not generate energy. Because of that, they can "produce" or deliver much more energy than contained in the energy it takes to run them.

This May Be Hard to Believe, But …

Not only is it possible to extract heat from air, water and earth, it is also possible to store heat energy in the ground. This is the most counter-intuitive concept the reader is likely to encounter this month. One would think that once heat is pumped down into the earth it would simply dissipate and could never be retrieved. Fortunately, such is not the case.

Geothermal storage is an extremely viable form of heat energy storage. Viable, not because of its efficiency, which probably will never rise above 50%, but viable because it is extremely cheap and reliable and can be installed virtually anywhere. And it never wears out. When attached to an array of solar hot water panels pumping heat into the ground during the summer, COPs of 30:1 can be achieved.

That translates to an "efficiency rating" of 3000%. More on this in a later chapter.

EROI: A Societally Critical Metric

The ratio of Energy Returned for Energy Invested was developed by Charles Hall, an American systems ecologist. He found that the well-being, and subsequently the behaviour, of fish could be directly equated to the energy they used in obtaining the energy they needed to both grow and reproduce.

Prof. Hall turned from fish biology to energy economics and then to biophysical economics by way of developing the EROI ratio. He then began to see how it was applicable to humanity's relationship with all energy flows, but most particularly, with those of fossil fuels and other energy commodities.

Our society leapt into industrialization and science on the back of increasing access to ever cheaper and more abundant supplies of energy. For centuries after coal began to do mechanical work, EROIs consistently increased and energy was applied to more and more tasks. It is safe to say that during those centuries, the energy cost of getting more energy was rarely a consideration as it amounted to such a small proportion of the overall cost (Fig. 5.4).

But the most dense and accessible (read "richest") reserves of energy have been largely depleted and now producing energy is more complex and energy intensive. As EROIs fall, the net energy yield of the process becomes a great deal more important. For instance, the EROI of ethanol produced in most northern regions is close to 1:1.

In other words, they yield barely as much energy as it takes to produce them and, in some cases, less. This is like planting 100 wheat seeds and harvesting a crop consisting of 100 wheat seeds. The non-viability of that process would be apparent to everyone, especially the farmers. But given the distortions of money metrics which include financial subsidies at every level, ethanol is indeed touted as a green alternative to fossil fuels in many regions where it clearly is not.

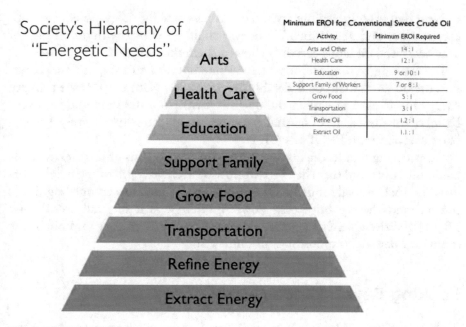

Society's Hierarchy of "Energetic Needs"

Minimum EROI for Conventional Sweet Crude Oil

Activity	Minimum EROI Required
Arts and Other	14 : 1
Health Care	12 : 1
Education	9 or 10 : 1
Support Family of Workers	7 or 8 : 1
Grow Food	5 : 1
Transportation	3 : 1
Refine Oil	1.2 : 1
Extract Oil	1.1 : 1

Arts
Health Care
Education
Support Family
Grow Food
Transportation
Refine Energy
Extract Energy

Fig. 5.4 Abundant energy is the base on which a sophisticated society grows. EROI is a prime development potential indicator

Looking Beyond Dollars

The biophysical economics metric of EROI would instantly expose many bio-fuels as being a waste of effort because their production consumes as much fossil fuel as they displace when they are burned. Biophysical metrics, i.e. measuring real physical material in physical units, cuts through the fog of money metrics which have, in many instances, misdirected environmental and energy policy.

Energy decisions have to be energy rational for the society, not just dollar rational for narrow interests.

Consider the example of the oil sands in Alberta. This is a huge resource amounting to 170 billion barrels of oil embedded in a sand layer stretching from east to west under central Alberta. There is enough recoverable oil in the oil sands to meet current US demand for 22 years. At $50 a barrel, the sands represent a gross revenue potential of $8.5 trillion. If developed over 100 years this would mean a contribution to the GDP of $21,000 every year for each of Alberta's four million residents.

Now that is an incredibly attractive business case for development for both corporations and government! Whoever could go wrong with an oil play of that size?? The oil sands certainly showed them the money.

But looking beyond the top line revenue figures and using different metrics, the oil sands looks considerably less attractive. Natural gas is the primary energy input used to extract the oil (bitumen) from the sand and for every barrel of oil equivalent of natural gas energy used, the output is only 3.5 or 4 barrels of oil for an EROI of <4:1.

Considering that in the glory years of virgin oil EROIs of those early fields could hit 100:1 and that the 2021 worldwide average is still in the neighbourhood of 16:1, the oil sands EROI of 3.5:1 looks much less compelling. Does the oil sands have a huge cash flow? Absolutely. Is it societally profitable? Highly questionable. Is it a good energetic bargain? Distinctly improbable. Is it environmentally responsible? Absolutely not.

Looking Beyond Today

And then there are the legacy costs. Once the oil has been extracted and the companies have shut down, how much will it cost to clean up the tailing ponds for which the oil sands are so infamous? To this point about 5% of the oil sands have been mined and the cleanup bill is between $30 billion and $130 billion. Oil companies have posted remediation security deposits of less than $2 billion.

The oil companies are adopting cleaner methods but the ultimate cleanup costs remain an open question. The oil business is subject to booms and busts and what happens to the liabilities after a few cycles which see companies with major cleanup liabilities going bankrupt? Taxpayers and their children can't leave town as easily as corporate liabilities can.

Beyond the dollars and cents, another consideration is the possibility that the oil sands as an "asset" might be "stranded" by a combination of electrification of the transportation sector and carbon emissions standards. Transportation is by far the largest consumer of oil and effective ghg reduction strategies to mitigate that pesky global warming/societal collapse threat would have a devastating impact on oil demand.

There is currently 5 times more verified reserves of fossil fuel sitting in the ground than can fit in humanity's remaining "carbon budget" – the amount of carbon we can emit before pushing global temperatures over the critical 2 °C level. But then, the world commercial economy always wants more cheap energy. Hence it is difficult to forecast the market for oil or its price.

One thing that is not difficult to do though is to predict that as long as the oil sands is being developed, it will consume a great deal of natural gas. Natural gas is the cleanest of the fossil fuels in carbon emissions terms and it also interfaces beautifully with a renewable energy grid system.

How so? Natural gas electric generating plants can react very quickly and this makes them ideal for complementing the variability of solar and wind renewable energy generation. The only other fast reacting source is hydro because coal and nuclear plants both take much longer to increase their output. Using a great deal of natural gas to produce oil from a low grade resource like the oil sands is the equivalent of spinning gold into lead.

EROI is a crucial biophysical metric and if it had been applied as a lens through which to evaluate the oil sands, the governments of Alberta and Canada might have been a little more sober in their assessment of the resource as a socially, fiscally and environmentally viable undertaking.

Net energy availability defines the limits of progress a society can achieve. EROI is an essential survival tool for a sophisticated society.

Declining Energy Richness

In a few short centuries, humans have already eaten most of the centre of the energy steak and are working outwards towards the bone and gristle. The period of rapidly increasing EROIs, which accompanied the early period of fossil fuel development, have given way to their rapid decline as the richest fields deplete (Fig. 5.5).

The new renewable sources harvest much more diffuse energy flows, rather than mine dense energy stocks and consequently require more energy input

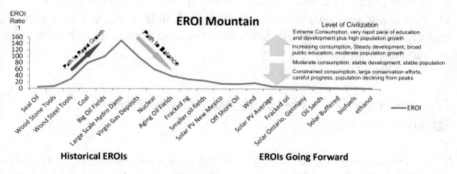

Fig. 5.5 EROIs increased during humanity's boom period but are now declining. Can we hang on to our gains with improved technology and efficiency?

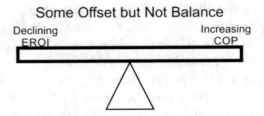

Some Offset but Not Balance

Fig. 5.6 Increasing efficiency is unlikely to offset declining energy availability. Overall final demand simply has to go down

for any energy produced. That said, the processes we use and the way we handle energy have both become much more efficient and the magnitude of this increase clearly offsets a good deal of the lost production efficiency.

Still, we will have to change our attitude from one of endless possibilities to one of well-considered steps toward sustainability. The new energy grids will be more complicated than those of the past. We will also have to be much more careful with the energy we consume and more diligent in the myriad ways in which we will be harvesting it (Fig. 5.6).

Unavoidable Demand: Location, Location, Location

A great deal of the energy and resources we consume is discretionary but some of it is absolutely necessary. This is demand that can't be avoided and is a function of how harsh average conditions are at a particular location, and what extremes it presents. How much energy does this environment require humans to expend to be able to feed, clothe and build shelter for themselves? The greater the energy and time required to supply these essentials, the lower will be the available surplus to develop tools and skills used to create a more sophisticated society. Location determines the level of accessible resources and the basic need for them.

The Energy Overview: How Do All these Factors Add Up?

It has been estimated that an EROI of 8:1 generates enough net disposable energy to maintain a moderately sophisticated society. But the sophisticated society we are transitioning to does not necessarily resemble our globalized consumer society of today. The 8:1 ratio is an average and like all averages it may not describe reality for any specific region. Also, EROI is a measurement of the ease of obtaining a single, albeit critical, input; energy. There are many other considerations which contribute to social sophistication such as the ready availability of other resources.

For instance, imagine an Inuit community before contact with Europeans. In this society, one seal hunter might be able to support a family of 5^2 so the EROI might be 5:1. If more seals became available with the same effort and there was a surplus of seal oil to the point an EROI of 8:1 was achieved, what would the surplus energy be applied to?

There were minimal natural resources to develop in the harsh Arctic environment, long range travel was arduous and dangerous while making a larger tent or Igloo could easily consume a great deal of the excess energy. In other words, the richness of the environment, and the demand the environment makes in order to survive, has a huge impact on the level of EROI that is required for any given standard of living. Compare the Inuit to their southern cousins, the Mayans.

An EROI of 8:1 or even 5:1 on the coast of Central America might allow an extremely sophisticated society to flourish. Neither heat nor elaborate clothing was required and travel to get food from the sea or forests was short. Large stands of timber are readily at hand and possibly some mineral deposits as well which would allow them to make better tools.

The Central Americans could build cities and roads with their surplus time and energy. The Inuit might be able to increase their population slightly or maybe extend their very short lifespans. But they would never be able to build cities in the north and they would never be able to put an Inuit on the moon.

Money: What It Isn't

There is energy and there are other biophysical metrics to measure resource stocks and flows. Then there is money which represents no environmental or social processes or health but can be used to measure the level of activity in the commercial economy and the relative value of goods and services in that marketplace.

---

[2] One hunter, one wife, two children and one senior, most likely the 45 year-old grandmother as grandad would probably die on the hunt. Seals are used to represent the array of fish and animals available to the Inuit.

Money is a very useful tool but it has unfortunately been miscast first as a measure of progress and subsequently as a goal. Money was created to make the mechanism of trade more fluid and efficient. Instead of trading four chickens for one arrowhead or eight tree trunks for a donkey, money, in forms ranging from seashells to gold to printed paper, has been used to represent commonly recognized value. That is the role it was designed to fill.

Money is a claim on real wealth: real goods and services. But money is not real wealth; it is merely a trading medium for the transfer of wealth.[3]

Money is used to measure the amount of commerce being conducted in the commercial economy. GDP – Gross Domestic Product - is the value of all of the transactions which are monetized. Transactions which are barter, say trading one thing for another or doing some type of work to help someone or to offset an expense like rent etc., aren't counted. The size of the GDP only represents transactions which involve money. Hence, the problem with using GDP as an indicator of social or environmental health becomes apparent.

GDP Depiction of a Bike

GDP does not describe real output or outcomes or value to people, it merely indicates the dollar value of transactions which take place i.e. cash flow.

Through the lens of GDP, cycling is bad for the commercial economy and so are:

- Good health
- Fair weather
- Long lasting products
- Low crime
- Recycling
- Low debt
- Buying local
- Stable population
- Short commutes
- Affordable housing
- Accessible nature

---

[3] For a more thorough explanation of money and its history, see: www.theperfectcurrency.org

Fig. 5.7 There is a clear difference between what is good for society and what is good for the commercial economy

In effect, efficiency and productivity are bad for the economy according to the GDP metric because it does not measure real output or beneficial impacts, but only the amount of paid activity (Fig. 5.7).

Picking the Right Tool Out of the Toolbox

We need to carefully select the right metrics and understand what they mean. We need to understand what we are measuring and what we are hoping to achieve.

For the first time in history, politicians of all stripes stood in front of large charts showing the growth of Covid-19 infections. These exponential curves gave an early warning of a looming crisis. They were the right tool. They illuminated the issue.

Being able to spot a trend is critical. The fact that 15 people died of Covid-19 on a Wednesday in Italian hospitals didn't worry scientists. What worried them was the fact that on Monday, eight people had died and on the previous Wednesday the first person had died. A curve which appears to be exponential will quickly get the attention of scientists because what starts out as an occasional drip and turns into a trickle will turn into a torrent extremely rapidly.

A doubling of deaths every 2 days means that in 20 days thousands will be dying every 24 h. Problems compound just the same way interest does.

Figure 5.8 shows a straight line projection of anticipated global temperatures in blue. The red line represents the actual measurements of global temperatures. These two lines diverge. Note the red line clearly shows evidence of accelerated temperature increase.

If the humanity's response to global warming is based on it being a linear growth event when it is in fact an exponential growth event, the implications for outcomes is immense. The increase in the rate of temperature has come to be known in the atmospheric science community, appropriately enough, as "**The Acceleration**".

We know the climate is warming globally but the latest analysis indicates that the rate of warming is increasing. This had been widely hypothesized but is it really happening? If so, what does that do to the strategy we have in place to deal with climate change as we understood it several years ago?

Warming has accelerated to ~0.25 °C for the most recent 2010–19 decade. Average decadal rate of warming prior to 2010 was ≤0.2 °C. The next 25 years are projected to warm at a rate of 0.25–0.35 °C per decade.[4]

What would have happened if we had believed that the growth rate of the Covid-19 virus was linear and instead of taking 20 days to increase 1000 times, the timing would have been years? Would we have responded as vigorously as we did? (Fig. 5.9)

---

[4]Xu, Y, Ramanathan V & Victor, DG 2018, 'Global warming will happen faster than we think', *Nature*, vol. 564, pp. 30–32; Tebaldi, C et al., 2020, 'Climate model projections from the Scenario Model Intercomparison Project (ScenarioMIP) of CMIP6', *Earth System Dynamics*, 16 September, pre-print.

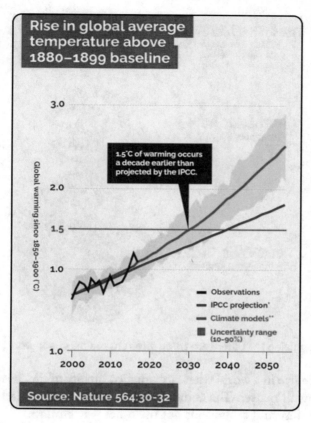

Fig. 5.8 Linear or exponential trend. We conflate them at our own peril. (Source: Nature 564:30–32)

The Folly of Choosing the Wrong Yardstick

Monetizing nature is at best a waste of energy and time. The great majority of natural processes take place outside of the commercial economy. In the past, we have attempted to price these as value to humanity or tried to rationalize the destruction of environmental systems based on the cost or benefit to the commercial marketplace. This simply misses the point that the commercial economy does not represent a full context forum for human or natural well-being.

For many governments and all businesses, money and the size and growth rate of the commercial economy, the GDP, represent both the immediate and the only truth.

The Immediate Truth
What business and
governments see

$ GDP $

The Popular Truth
What people live

Social Cohesion

The Greater Truth
How the planet functions

Biophysical Economics

Fig. 5.9 Recognizing that there are three different realms in our world is vital

But people live in a world where job quality, quality of life, health and family well-being all matter. The commercial economy functions to support the society's structure and aspirations not the other way around.

The natural world is the basis for all life and the human activity it can sustain should be measured in the physical units which best represent its complex stocks and flows.

The commercial economy must function as a mechanism in support of society and the entire human enterprise[5] must make itself compatible with the natural world (Fig. 5.10).

Monetizing Nature

One can think of Mother Nature as the womb inside which humans exist. Attempting to put a monetary price on essentials like this is a wasted exercise. Survival and social cohesion are targets which must be met regardless of the

---

[5] Bill Rees, https://www.youtube.com/watch?v=YnEXEIp5vB8&t=4341s

Fig. 5.10   We need a full perspective to properly setting our priorities

impact on the commercial economy. The commercial economy operates within a society which in turn operates within the biosphere.

How much will it cost to mitigate climate change? How much will it cost to save the Amazon rain forest or the health of our oceans? It doesn't matter; just know these are jobs which must be done if our society is to remain viable. Money can be used to allocate resources for the job within the commercial economy but money must never be used to determine human priorities or social or environmental well-being.

Budgets: Energy

We need to represent our consumption of energy in all its forms by way of per capita daily or annual budgets. Once we know how much we are spending, we can begin to take that kWh per capita figure and use it as a yardstick to measure our progress towards lower consumption and the shift to low carbon, renewable energy sources.

Budgets: Carbon

There is the energy budget and then there is the carbon budget. The carbon budget is much more complex because it includes not just the carbon content of the energy people consume directly but also the carbon used in making

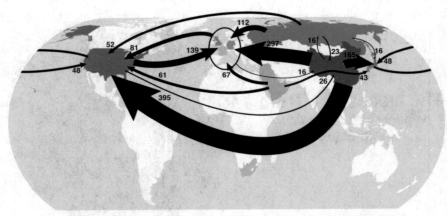

Consumption-based accounting of CO2 emissions Steven J. Davis1 and Ken Caldeira

Fig. 5.11 Carbon emissions are embedded in everything we eat and consume whether produced domestically or imported

(embedded carbon) the material goods and food consumed in the country and perhaps produced elsewhere (Fig. 5.11).

A widget made in a country with a very clean energy mix, consisting of wind, solar and hydro, has much lower "embedded" carbon than the same widget made in China. There, a very high percentage of energy is supplied by coal. This will be covered more in a later chapter but for an individual in the developed world, globalized supply chains are a large contributor to the size of their carbon footprint.

In general, locally produced goods will contain much lower levels of embedded carbon. Buying local, as much as it is an anathema to globalism promoters, is simply the easiest way to make your buying decisions the cleanest they can be.

If you are trying to come to grips with the carbon footprint of your food, there are numerous carbon calculators to consult. The ones created by groups in your region would give the most reliable estimate for food consumed in your geographic area. The referenced video[6] gives an excellent overview of the issues and should enable you to better judge which choice is the lower carbon one.

---

[6] https://www.youtube.com/watch?v=jk_YGNzBwUo

Budgets: Renewable and Non-renewable Resources

We need to shift our consumption of non-renewable resources to resources which are either renewable or which allow very high rates of re-cycling.

There is a rapidly growing network of recycling apps and monitoring organizations to inform individuals and planners on the best choices and trends in recycling. If it can't be recycled, don't buy it or plan on keeping it for many decades.

Renewable Energy: All Around Us but Difficult to Grasp

Renewable energy sources are ubiquitous but diffuse. We are surrounded by energy flows but much larger areas must be covered by solar and wind infrastructure than is currently covered by fossil fuel and nuclear mining and processing facilities. Keep EROI and net energy in mind when examining energy options.

- A solar farm outputs 5 W/m^2 at northern latitudes but 10 W/m^2 in the US SW
- Nuclear plants produce 1000 W/m^2
- Wind power produces 2.5 W/m^2

Energy **content** means the total energy potential of something. It does not mean the **net usable energy** available to do work as the above comparison between the internal combustion engine and an electric motor illustrated.

- 1 cu ft. of natural gas = 0.3 kWh
- 1 cu m of natural gas = 9 kWh
- 1 L of gasoline = 9 kWh
- 1 US gallon of gasoline = 33 kWh
- 1 Imperial gallon of gasoline = 37 kWh
- 1 barrel of oil = 1700 kWh

KWh is likely to be increasingly used as the standard measure of energy consumption and production. It is ideal to represent the energy transmitted through the grid, produced by solar panels or wind generators and consumed

Table 5.1 Conventional combustion fuels expressed in their electrical energy equivalent

Energy source	Unit	Energy content kilo Watt hours
Electricity	1 kilowatt-hour	1 kWh
Butane	1 cubic foot (cu. ft.)	0.9
Coal	1 ton	8206.3
Crude oil	1 barrel - 42 gallons	1699.9
Fuel oil no.1	1 gallon	40.3
Fuel oil no.6	1 gallon	44.7
Diesel fuel	1 gallon	40.7
Gasoline	1 gallon	36.3
Natural gas	1 cubic foot (cu. ft.)	0.3
Heating oil	1 gallon	40.7
Kerosene	1 gallon	39.6
Pellets	1 ton	4835.9
Propane LPG (liquid petroleum gas)	1 gallon	26.8
Propane gas 60oF	1 cubic foot (cu. ft.)	0.7
Residual fuel oil 1	1 barrel - 42 gallons	1842.6
Wood - air dried	1 cord	5861.7
Wood - air dried	1 pound	2.3

in a car or house. It might be new to many but it tremendously simplifies energy measurement and budgeting. A single metric allows rational investment in energy production and consumption to begin in earnest.

For fossil fuels and wood which involve combustion processes,[7] the actual work performed ranges from about 30% to 80% of the energy content (Table 5.1).

Precision Versus Feel

Knowing you are moving in the right direction is more important than knowing precisely where you are in real time. Certainly accurate monitoring is important but it is also very easy to drown in data. Let scientists do that; they are paid to separate the forest from the pine needles. We just need to know we are on the right path.

There is an immense amount of information available for every aspect of the issues touched on in this book. Arriving at a basic understanding of the mechanics of energy generation and conservation and getting a feel for how they affect your life is far more important than making decisions accurate to four decimal places.

---

[7] Energy Content, https://www.engineeringtoolbox.com/energy-content-d_868.html

Will an efficient EV, that meets your needs be a big step towards reducing your carbon footprint? Yes. Will replacing large SUVs and trucks with electric Hummers save the planet? No.

And by saving the planet, we mean maintaining earth's environmental systems in a healthy enough state to sustain moderately sophisticated human societies. The consumer society is beyond saving but the sustainable society, that can continue to learn and progress, is not.

We need to understand how best to work with what Mother Nature has to offer us and neither simple nor complex math is enough to provide that. We need to be able to step back and once again develop a feel for what is right for us and the planet, just like our pre-fossil fuel ancestors.

6

How to Keep the Planet Human-Friendly

Worldly Problems Require a Biophysical Outlook

A human-friendly earth demands that conditions on the planet be made not only suitable for subsistence, but suitable as well for the on-going prosperity of human societies. The term "prosperous" is not held here to be the continuance of reckless and vainglorious consumption. Rather than expanding version our current consumer society, we need to stabilize and become advanced enough to continue to increase the stock of human knowledge while supporting high levels of human well-being.

This means that humanity will have to meet a number of conditions, among them:

- Assuring the continued existence of as many species as possible by:

  - Reducing the stress we place on their habitat
  - By not displacing them any further
  - Allowing large areas to return to nature

- Assuring the recovery of the world's major biosystems to the point they are no longer under threat of decline.
- Decarbonizing our activities to limit the danger of global warming to levels which will avoid social disruption.
- Reduce our consumption of raw materials through, re-using, recycling and much lower consumption levels to allow the lifespans of our recoverable mineral reserves to be extended by centuries.

J. E. Meyer, *The Post-Pandemic World*, https://doi.org/10.1007/978-3-030-91782-1_6

The scale of the human enterprise on earth has to dramatically shrink through the combined effects of reduced consumption and increased efficiency. Certainly our population can grow no larger and should be encouraged to go into slow decline. But humanity actually is growing larger as lagging nations still show fertility levels above replacement levels. That means consumption levels have to fall in responsible nations to counter the population increase in the irresponsible nations.

Loving Nature, Loving Ourselves

If we are good to ourselves and live healthy lifestyles, we will most likely impose much less stress on the planet. But if we continue to act like fearful, desperate consumers, the prognosis for the planet is not good.

During the pandemic, there was a stampede out of dense urban cores to the space and clean air of more natural surroundings. If they had the money, people sought out healthier, less crowded areas to ride out the crisis. Many committed to this relocation by selling their properties in-town and buying country residences. As it turned out, healthy and natural surroundings simply offered a better lifestyle and a sizable portion of the population now has this front of mind.

Our Numbers

As hunter-gatherers, we used to limit our numbers to match the carrying capacity of the land. When we began to exert control over the land through large-scale farming, we began to think of nature as a resource ATM. And now commercial economics assumes the loving mother on the other side of the wall would never cut us off. But Mother Nature has indeed repeatedly and predictably cut support for human populations once they began to have a significant deleterious impact on the health of their local environments.

Population cycles of Grow – Deplete – Collapse – Repeat, is a recurring process human societies experience over 250–400 year time spans.[1] Breaking the boom/crash pattern and establishing social policy with intergenerational perspective will require us to fundamentally change the metrics and accounting systems we use as well as the goals we set. Personal, corporate and

---

[1] https://www.youtube.com/watch?v=Yw6I_PIdTIQ&t=10s

governmental responsibility will have to be expanded to encompass the greater long term good.

Words from Great Thinkers (1)

Garret Hardin from his "Tragedy of the Commons": "The population problem has no technical solution; it requires a fundamental extension in morality."

"It is fair to say that most people who anguish over the population problem are trying to find a way to avoid the evils of overpopulation without relinquishing any of the privileges they now enjoy. They think that farming the seas or developing new strains of wheat will solve the problem – technologically. I try to show here that the solution they seek cannot be found. The population problem cannot be solved in a technical way … "

"The tragedy of the commons develops in this way. Picture a pasture open to all. It is to be expected that each herdsman will try to keep as many cattle as possible on the commons. Such an arrangement may work reasonably satisfactorily for centuries because tribal wars, poaching, and disease keep the numbers of both man and beast well below the carrying capacity of the land. Finally, however, comes the day of reckoning, that is, the day when the long-desired goal of social stability becomes a reality. At this point, the inherent logic of the commons remorselessly generates tragedy.

As a rational being, each herdsman seeks to maximize his gain. Explicitly or implicitly, more or less consciously, he asks, "What is the utility to me of adding one more animal to my herd?" This utility has one negative and one positive component.

1. *The positive component is a function of the increment of one animal. Since the herdsman receives all the proceeds from the sale of the additional animal, the positive utility is nearly +1.*
2. *The negative component is a function of the additional overgrazing created by one more animal. Since, however, the effects of overgrazing are shared by all the herdsmen, the negative utility for any particular decision-making herdsman is only a fraction of 1."*

"Adding together the component partial utilities, the rational herdsman concludes that the only sensible course for him to pursue is to add another animal to his herd. And another; and another… But this is the conclusion reached by each and every rational herdsman sharing a commons. Therein is the tragedy. Each man is locked into a system that compels him to increase his herd without limit – in a world that is limited. Ruin is the destination toward which all men rush, each pursuing his own best interest in a society that believes in the freedom of the commons. Freedom in a commons brings ruin to all."

"Freedom without responsibility in equal measure is the path to decline."

Similarly, the balance between freedom and responsibility seems to be a concept that has eluded anti-maskers and anti-vaxxers during the Covid crisis.

Economically logical decisions by individuals do not necessarily add up to actions which make societal or environmental sense. Self-centredness, as we have seen during the pandemic, has led to far worse outcomes for most people and certainly for the society as a whole.

Global Commons

The inevitable Black Friday brawls in Aisle 7 in Walmart for 50″ TVs at 70% off are but a preview of international conflict as countries seek to assure the supply of dwindling critical energy, water and raw material resources.

China has a clear policy of establishing direct control over a wide range of critical commodities in countries far from its own borders. This is not participation in a market system; this is using a supposed market system to further a national strategy of direct ownership and control of the resources themselves as well as their means of distribution.

China is positioning itself to maintain social stability for its people through the stability and resilience of its energy and raw material supply systems. If the globalism mantra of economic elites in western democracies carried any weight with China's leadership, they would simply rely on market forces to provide their long term requirements.

China has exploited the globalism movement to the fullest in building themselves into the world's dominant manufacturer and technological powerhouse in record time. But they aren't depending on market mechanisms for resource security; they are depending on direct control of mines, farms, ports and rail systems. As well, they are employing the old economic imperialist mechanism of building infrastructure in less developed countries and owning the debt these projects create. Unlike their western predecessors however, they are doing all of this without firing a shot.

When other counties find themselves coming up short in critical supplies, how will they react? If they have large military forces, there will be strong incentives to take what they need to avoid domestic social disruption. When elites declare war on someone outside their borders, pent-up domestic dissatisfaction transfers from the elites to the foreign enemy. This is the process that has played out in larger and smaller scales throughout history.

The Entire Human Enterprise

Technical competence is one thing, but cultural understanding i.e. awareness built into the culture, is something else entirely. Unless awareness of biophysical limits once again becomes infused into our culture and policy making, we will find it impossible to accept or mitigate the coming biophysical shocks.

Our awareness needs to be focused on the quality of life rather than on the "spending more, living better" theme. Currently, to policy makers, it matters not whether we are the Walmart or the Gucci nation; it only matters that consumption is a widespread obsession. This simply has to change because there is no technical solution for marrying this lifestyle with the realities of the natural world (Fig. 6.1).

If we aren't going to look after earth's natural systems, we are going to have to create an earth on which all systems are man-made and regulated. Our current policy makers appear to believe this is possible.

Infinite Humanity

There are those who feel that not only can humanity expand both its numbers and consumption for eternity but must do so to avoid economic collapse. This concept is put forward by a small number of commercial economists tied to the growth dependent sectors of media, finance and development. Certainly the cessation of building more houses and shopping malls along with the debt and asset inflation they generate would negatively impact those sectors but lower debt and more affordable housing would be a boon to the rest of society.

Fig. 6.1 Will we choose to live in the natural world or shopping malls in space?

Needless to say, the promotion of the mantra of infinite growth on a finite planet is something physical scientists find laughable. But their mirth is tempered somewhat by the realization that some governments, including those in Australia, Brazil, Canada and perhaps the USA are still committed to the concept.

Signs of a Very Sick Planet

A full understanding of the planet means that we have to recognize that global warming will create a large number of complex impacts far beyond mere temperature variations. Who will really care when Miami partially disappears every high tide if mid-west food production has fallen to zero due to drought?

Technical paper 6 from the IPCC on Climate Change and Water notes that observed warming over several decades has been linked to changes in the large-scale hydrological cycle. The specific impacts include:

- increasing atmospheric water vapour content
- changing precipitation patterns
- intensity and extremes
- reduced snow cover
- widespread melting of ice
- changes in soil moisture and runoff

Globally, the area of land classified as very dry has more than doubled since the 1970s. These trends are accelerating.

What Have We Done So Far?

Environmental damage done by humans probably goes back further than the initial colonization of Australia some 50,000 years ago. There we obliterated the large herbivore populations and perhaps turned the continent into the mostly desert region it is today.[2]

Environmental destruction in Scotland 4000 years ago is perhaps a better documented illustration of what happens when a growing population hits sustainable limits; they eat their seed grain. "Seed Grain" could take the form of destruction of the soil, depletion of fisheries or, as in the case of Scotland,

---

[2] The Future Eaters, Tim Flannery.

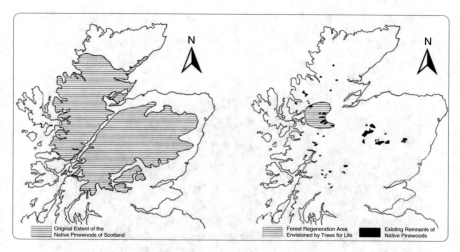

Fig. 6.2 Deforestation is often human caused but it can also be human remediated

the over-cutting of their forests which involved the reduction of native pinewood forests by 98%.

Forests and eco-systems can be re-built but it takes time and dedication as outlined in the presentation by Alan Watson Featherstone called "Restoring the ancient Caledonian Forest" (Fig. 6.2).[3]

Killing off species[4] is both a by-product and a clear intention of human efforts to terraform the planet for our own short-term benefit. Loss of biodiversity, according to Jared Diamond, is caused by the "evil quartet" of overkill, habitat destruction, secondary extinctions, and introduced species. Factors identified by Edward Wilson are habitat destruction, climate change, invasive species, pollution, human overpopulation, and over-harvesting. The common theme in both analyses is that we are simply using more than the earth can sustain (Fig. 6.3).

Not everyone will look as happy as this group of researchers when they discover a Burmese Python in their back yard or swimming pool. These reptiles are fanning out over Florida and reducing the populations of native species by as much as 95%. As the climate warms, they are heading north.

"Honey, have you seen the dog this morning?"

Less intimidating but still very harmful species are Asian carp, and species of mosquitoes and ticks which carry tropical diseases northward.

---

[3] https://www.youtube.com/watch?v=nAGHUkby2Is
[4] https://www.conserve-energy-future.com/what-is-biodiversity.php

Fig. 6.3 When removing a garden pest of this size it is best to enlist the help of people who you really trust

Species can also be threatened by genetic pollution – uncontrolled hybridization and gene swamping. For instance, abundant species can interbreed with rare species, thus causing swamping of the gene pool.

Overexploitation is caused by activities such as overfishing, overhunting, excessive logging and illegal trade of wildlife. Over 25% of global fisheries are being fished at unsustainable levels.

Changes in climates and global temperatures directly impact physical environmental factors essential for sustainable habitat. For example, if the present rate of global warming continues, habitats from mountain to coral reefs, which are biodiversity hotspots, will disappear in 20–40 years.

The wildlife in the mountain regions which require the cool temperatures of high elevations such as the rock rabbit and mountain gorillas may see their habitat "sink" into the rising ocean of warm air. 10% of all species might go extinct by 2050 (Fig. 6.4).

The various forms of pollution, including water, soil, and air, pose a serious threat to animal and plant habitats due to the release of toxic substances and chemicals.

Overpopulation has driven continued encroachment into frontier forests, heightened pollution, and destruction of natural ecosystems that have considerably contributed to the mass extinction of species (Fig. 6.5).

The lone orangutan in this tree is barely visible and he and his band will disappear completely as their habitat is displaced by farms and urban sprawl.

Fig. 6.4   A lone tree of a once great forest with a lone survivor

Fig. 6.5   A species fades from history

Orangutans are just one more species fading into history to make room for the expanding human enterprise.[5,6]

According to the, IPCC (Intergovernmental Panel on Climate Change and Biodiversity): "At the global level, human activities have caused and will continue to cause a loss in biodiversity through, land-use and land-cover change; soil and water pollution and degradation (including desertification), and air pollution; diversion of water to intensively managed ecosystems and urban

---

[5] http://www.nwf.org/Wildlife/Wildlife-Conservation/Biodiversity.aspx
[6] https://www.nationalgeographic.org/encyclopedia/biodiversity

systems; habitat fragmentation; selective exploitation of species; the introduction of non-native species; and stratospheric ozone depletion."

And the IPCC goes on to describe a continued slow-motion environmental train wreck. "The general effect of projected human-induced climate change is that the habitats of many species will move poleward or upward from their current locations. They will attempt to migrate at different rates through fragmented landscapes, and ecosystems dominated by long-lived species (e.g., long-lived trees) will often be slow to show evidence of change. Thus, the composition of most current ecosystems is likely to change, as species that make up an ecosystem are unlikely to shift together."

The negative changes which will occur for our biosystems by the dynamics we have unleashed are unlikely to have a positive impact on the welfare of humanity.

Signs of a Sick Humanity

There isn't any need here to go over the distress alarms sounding in many countries on an on-going basis. Daily news of these symptoms of social stress is non-stop. The question is; are the levels of strife staying constant or growing worse?

- civil disruption
- inequality
- debt
- migration

War is the final expression of social stress. but war, despite the proliferation of arms worldwide, has been less common in recent decades than in many periods of the past. Are we mitigating the effects of stress better than we have through history? Are our weapons now too potent to even consider a large scale war or have we been experiencing an interlude of high life satisfaction among the planet's now eight billion people? How long the general peace will last is perhaps directly tied to how well we manage the increase in biophysical dynamics.

Migration is the clearest indicator of extreme social stress before war actually breaks out. Ifeyinwa Ugochukwu,[7] CEO of the Tony Elumelu Foundation,

---

[7] https://www.euronews.com/2020/01/22/africans-wouldn-t-risk-trip-to-europe-if-they-had-economic-opportunities-at-home-says-ceo

commented "The only reason that Africans risk their lives on the Mediterranean Sea and go into unknown areas is they are looking for economic opportunity …. if they have it in their home, they wouldn't go elsewhere looking for it."

Humanity to this point does not seem driven to solve the problems it has where they originate but rather seems to content to allow them to spread themselves around.

Not Planet Friendly

Urban growth has sprawled around a tiny indigenous settlement in San Paulo, Brazil.[8] "I never wanted to live there, but the city insists on coming to us," tribe member Balbina Terue said. "I don't see why people have to destroy the environment just to live here."

According to the developer dogma which pervades government in some countries, this is progress. Population growth driven urban sprawl is the full package when it comes to social and environmental negative impacts (Fig. 6.6).

Fig. 6.6   Can urban sprawl and the natural world co-exist? No

---

[8] https://nypost.com/2020/02/05/tiny-indigenous-land-highlights-brazils-environmental-woes/

This trend plays out in countries which are still promoting high population growth from Brazil to Canada to Australia. Urban development and the debt and asset inflation which go with it are the largest drivers of wealth for the very small elites who run those countries.

Planet Friendly

Conversely, secure in the certainty that human activities must become planet friendly, Japan has created human-centred smart cities designed to enhancing well-being in balance with nature.[9]

The Japanese call it Society 5.0 and it is being applied in the towns of Aizuwakamatsu and Arao which have adopted the latest technologies to improve people's well-being. It's a practical application of AI to help people live healthier, happier and longer lives. This is similar in thinking to the Bhutanese Index of Happiness metrics but applied also to direct involvement with each participating individual.

City layouts and environments will be modified to enhance well-being and access to nature but these programs also cut close to state monitoring and subtle direction of individual habits. Is it any worse than having private corporations dictate personal preferences and social priorities? Not likely, as long as the process remains transparent. Transparency, however, has never been a strong point for any type of human elite.

What Does Sustainability Look like?

Sustainability is a concept embedded in every hunter-gatherer culture going back through time but it is something few in a technically advanced society can describe in concrete terms. For humans in the twenty-first century, sustainability is a theory. For most of our ancestors, sustainability was a way of life. An Amerindian chief in the US north-west in the 1800s could see the conflict between the rapacious nature of European settlers and his own way of life which he knew was being extinguished. Here are his thoughts on the short-term, self-centred thinking of a rapidly expanding colonial population desperate for new lands to exploit.

---

[9] https://www.euronews.com/2021/01/18/japan-s-human-centred-smart-cities-enhancing-well-being

Words from Great Thinkers (2)

Chief Sealth

"Every part of the earth is sacred to my people. Every shining pine needle, every sandy shore, every mist in the dark woods, every clearing and humming insect is holy in the memory and experience of my people.

The white man is a stranger who comes in the night and takes from the land whatever he needs. The earth is not his brother but his enemy and when he has conquered it he moves on. He leaves his fathers' graves and his children's birthright is forgotten.

All Things share the same breath - the beasts, the trees, the man. The white man does not seem to notice the air he breathes. Like a man dying for many days, he is numb to the stench ... What is man without the beasts? If all the beasts were gone, men would die from great loneliness of spirit, for whatever happens to the beast also happens to man.

All things are connected. Whatever befalls the earth befalls the sons of earth ... The whites too shall pass - perhaps sooner than other tribes.

Continue to contaminate your own bed, and you will one night suffocate in your own waste. When the buffalo are all slaughtered, the wild horses all tamed, the secret corners of the forest heavy with the scent of many men, and the view of the ripe hills blotted by talking wires, where is the thicket? Gone.

Where is the eagle? Gone.

And what is it to say good-by to the swift pony and the hunt, the end of living and the beginning of survival."

Conversely, for the post-consumer economy, the end of growth means the beginning of sustainability.

Our Lifestyle; Yes, There Will Be Changes

Instead of owning everything, rent when you need it. Instead of buying a vehicle which can do everything except be efficient, buy a vehicle that will do what you need 95% of the time and rent or hire the last 5%.

We have to change the first world vision of getting into a vehicle as a prerequisite for every activity. As well, vainglorious consumption can no longer be the way to assure someone of high social status.

A short list of ways to assure that your society's lifestyle unfolds in a safe and enjoyable manner would include going electric, developing daily per capita energy budgets, making your house energy positive. Thinking in terms beyond money metrics allows an individual to truly understand the biophysical systems which underpin our prosperity and survival.

Go Electric Because It Is Continually Improving

Electrical systems will continue to displace fossil fuel ones until fossil fueled devices become antiques, shunned by all except hobbyists and collectors. Electrical grids are now constantly lowering their carbon emissions per kWh of production so any decision to replace a fossil fuel device with an electric one will become an even better carbon decision going forward (Fig. 6.7 and Table 6.1).

"It's just a back-up system
for your pacemaker."

Fig. 6.7 Jim Unger's Herman serves up the new sustainable world

Table 6.1 Carbon emissions by energy source

US grid source[a]	CO_2 emissions Grams/kWh
Coal	900
Natural gas	450
Petroleum	850
Wind	5
Solar	8
Nuclear	4

[a]https://www.eia.gov/tools/faqs/faq.php?id=74&t=11

Carbon Versus CO_2 Weights

Why is the weight of CO_2 emitted greater than the weight of the carbon in the gasoline (or coal, ng) itself? Carbon dioxide is a molecule made up of 1 carbon atom and 2 oxygen atoms. The atomic mass of a carbon atom is 12 and the mass of an oxygen atom is 16 so the total weight is 12 + 16 + 16 = 44. That is why 1 L of gasoline, weighing 1 kg, when burned, can produce 2.3 kg of carbon dioxide. Gasoline is about 63% carbon.

Different countries have different fuel mixes and these mixes are changing. Becoming aware of the carbon content of your local power provider is a good way to make rational carbon decisions.

According to a study by Michaja Pehl,[10] fossil fuel generation, despite having carbon capture by 2050 including coal, ng and petroleum, will still have carbon emissions which are 5 times higher than those needed to meet even the 2 °C degree warming target. Other study highlights include:

- lifecycle emissions from fossil fuel plants even with carbon capture and sequestration range from 78 to 110 g of CO_2 per kWh
- lifecycle emissions from nuclear, wind, and solar PV range from 3.5 to 12 g of CO_2 per kWh
- **The global target for 2 °C warming is 15 g/kWh**
- coal with carbon capture has an EROI of 9:1, nuclear is 20:1, wind is 44:1
- Expected EROI for solar PV is 26:1 by 2050 having improved from the 2010 level of 10:1
- Hydro dam carbon emissions are largely due to rotting organic matter in the flooded area so dams should not be in warm climates with shallow reservoirs.
- Geography counts! - The best solar pv rating is 3 g CO_2/kWh and the worst is 21 g CO_2/kWh

---

[10] Understanding future emissions from low-carbon power systems by integration of life-cycle assessment and integrated energy modelling Nature Energy, Vol 2, December 2017.

Table 6.2 ICE vs EV by electrical source

Energy source	Energy use	Energy use/km	CO_2 emissions	CO_2 emissions Grams/kWh
ICE car gasoline	8 L/100 km	0.08 L/km	18.4 kg/100 km	184 g/km
EV car wind	18 kWh/100 km	0.18 kWh/km	5 g/kWh	0.9 g/km
EV car coal	18 kWh/100 km	0.18 kWh/km	900 grams/kWh	162 g/km

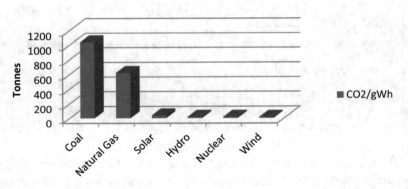

Fig. 6.8 Electricity carbon dioxide emissions by source fuel

Below, the CO_2 emissions of an electric vehicle change radically depending on the source of electricity (Table 6.2).

Here is a comparison of the Lifecycle Emissions for Various Methods of Energy Production (Asthma Society of Canada and Bruce Power, 2014). Different regions have different numbers but they all point in the same direction. Renewables are the only way forward to carbon emission reductions sufficient to give humanity a chance to avoid the worst impacts of climate change (Fig. 6.8).

The electrical grids of different countries use different mixes of electricity generation. Some regions like Norway, Sweden and Quebec and Manitoba have topographies which feature a great deal of water running off cliffs, facilitating large hydro dam construction. Other regions like Australia and Belarus, with large coal reserves and entrenched elites post vastly higher emissions (Fig. 6.9).

In other words, if you live in Australia, you shouldn't bother buying an EV to reduce your carbon footprint unless you will be mostly charging using the solar panels on your own roof (which most people will actually end up doing).

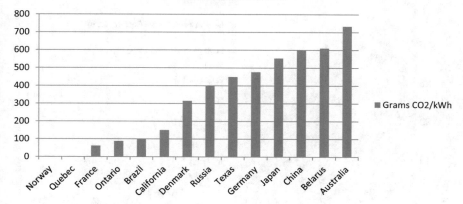

Fig. 6.9   Electricity is great but how clean are the grids?

Despite active government campaigns *AGAINST* EVs, sun washed Australia is changing. Much to the horror of the very powerful coal lobby, from 2011 to 2017 the CO_2/kWh carbon intensity of its economy has dropped from 999 to 733 and this trend will surely continue as it is in most parts of the world.

Recycling and Efficiency Are Only Way Forward (Fig. 6.10)

Humans habitually exploit the richest and lowest cost resource bodies first, and move to lower grade reserves as the richest bodies play out. The lower the ore grade, the more waste is generated and the more energy is required to remove the valuable minerals. Above, we see the gold ore grade of close to 50 g per ton in 1854 declining to 2 g per ton in 2012. This means 25 times more effort – energy, labour and technology - has to go into extracting the gold today than was required 170 years ago.

This process applies just as well to our farmlands as the current level of crop production is dependent on large inputs of fossil fuel energy and fertilizer.

Depletion is a trend that applies to all of the minerals on which humans depend. Our future turns on how well we can adopt the culture of conservation, recycling and efficiency. Lassoing asteroids in space and towing them back for processing on earth is an energetic impossibility just as is shooting our burgeoning population into space when earth becomes uninhabitable.

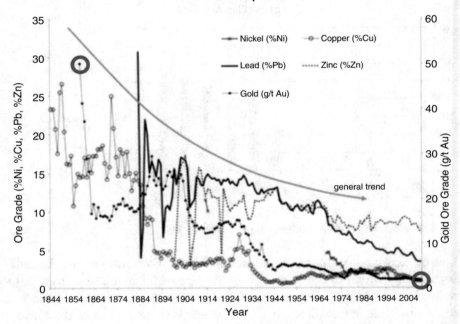

Fig. 6.10 More effort to extract less mineral yield. The easy reserves have been largely depleted

Structured for Success: Possibly

The Aging Trend Is Good

For those whose interests are embedded in the continual increase in consumption and asset inflation, aging of the population is a disaster. An older, stable population does not require more housing and certainly won't be paying more for what they need less. Seniors don't need the latest gadget or ego boost and they don't use as much, which means consumption will decline. Products are used longer, more gently and replaced far less often.

The aging trend is a planet-wide population transition which began in the eighteenth century when people began to live longer and have fewer children. This trend will stabilize with the end of population growth and with a higher percentage of seniors than ever before in human history. That is the demographic profile that societies from this point onward will be dealing with as the population boom era winds down.

Social Coherence

The renewable energy transition will help most people in many ways but it will also disrupt sizable groups and hurt some people. To maintain faith in institutions and social stability, it is necessary to make sure those suffering loss are not crushed by change. Covid-19 in countries which responded poorly saw their most vulnerable, as well as many of the middle class, hit the hardest.

In the USA, millions of people suffered income losses hindering their ability to pay for medical care or even rent. The tens of millions of Covid-19 cases caused a spike in health care costs at a time when incomes declined for many. Countries with better developed social safety nets will distribute the costs more broadly and avoid building hopelessness among a large portion of the public.

Covid-19 is a multi-year event but climate change and resource depletion challenges will extend over decades and possibly centuries. The short-term nature of the pandemic is unlikely to bring governments down but if the disparities generated during this crisis were extended over decades, fundamental changes in governance would likely occur.

Science Is Standing Up

Science is beginning to grow into the clear voids of public policy. Scientists and those groups advocating for science based policy are becoming much more aware of the shortcomings of the commercial economics based national policy system.

Information and modelling are a necessary prerequisite for the development of productive social policy and scientists are becoming painfully aware that most government policy is not anchored to biophysical reality. They are beginning to speak more loudly.

Awareness Is Growing

Although countries have been very slow to enact effective measures to deal with broad environmental issues, there is now a very high level of awareness in most governments that these issues are a huge threat and must be dealt with.

Similarly, the general public is much more aware of the need to reduce stress on the planet. As well, the in-effectiveness of the denial or half-measures approach has been fully exposed by the pandemic experience.

The School of Hard Knocks Pays Dividends

Given the spread and mutation capacity of the Covid-19 virus, it may take many years to be truly rid of it. But the lessons we have learned from it can pay off many times over in the context of preparing ourselves to deal with the changes required by climate change and resource exhaustion.

Maybe, in a way, Covid-19 is actually forming part of our societal immune system, teaching us what we have to do to protect ourselves.

> *The immune system is not a wall; it is an intelligence system that learns how to evaluate a threat and teaches the body how to counter it.*
> D. Verdon

Seeking Perspective and the Resurrection of a Healthy National Conversation

Market forces in the commercial world will not deliver favourable social outcomes for the large biophysical threats any more than they delivered an effective response to the Covid-19 pandemic.

Market incentives were used to a limited degree as part of the pandemic response but the goals and larger mechanisms were determined by government priorities. In a great many instances, these flew directly in the face of market guidance because the market is a fast reaction money/cash flow mechanism with no foresight and no social or environmental consciousness.

Governments however, are responsible for public welfare and environmental health over decades. The commercial market and its GDP metric were never designed to replace overarching national policy making.

Some governments executed their responsibilities very effectively and others didn't. But those who chose to let the commercial market determine national responses suffered the largest death rates and, ironically, the largest economic losses.

Clearly we are beginning to think in terms of interdependent mechanisms and relationships. Stewardship of the earth rather than exploitation is now on the agenda of most nations although exploitation still holds all of the money and most of the power. As much as national conversations have collapsed into polarized, often bitter, partisanship, we must not lose sight of the fact that a respectful national conversation is critical to long term prosperity. Our common ground needs to be literally what goes on under our feet and in the air around us.

A Eulogy to Truth: Long Live Honesty

The truth is dead - long live honesty
  Entailing honest accounts and holding accounts honest
  Science teaches us skepticism - Entailing multiple lines of evidence for reliable knowledge.
  Complexity teaches us relative perspectives - Entailing multiple ways of reasoning for relevant wisdom
  Collective wisdom emerges in our institutions of conversation
  Entailing good faith speaker-hearers - with honest accounting
  Entangling complex reasonings - For adaptive evolving
John Verdon describes the tenets of a healthy national conversation.

Knowledge and respect will move us away from the demand that our leaders know all and must deliver anything asked of them or that scientists be able to offer perfect theories and certain outcomes.

We need to enable honest social chemistry by understanding that there is no objective way to reason and that we can't know all variables in any long term unfolding of events. All we have is partial knowledge and innumerable ways of reasoning.

Absolute certainty can only be achieved in an environment of absolute ignorance.

If we can recognize the good faith of other peoples' experiences and evidence then maybe we can be in a better position to listen, to hear an honest account and in turn offer our own honest account. Together richer understanding emerges.

Achieving the wisdom to flourish and evolve requires transparent methods to ensure good-faith participants. The national conversation needs to be open to all evidence that is held to be accountable and honest. But more – we need to accept the inevitable uncertainty that is deeply entangled in all life and evolution.

Everyone who reads this book will become, to a lesser or greater degree, an energy thinker. They will still be very conscious of money but they will also be aware that money actually represents very little beyond the commercial marketplace and that the world is very much bigger than the market.

It is important to know what your daily energy budget is and its carbon content. Try and learn where your energy and material goods come from and make basic decisions which can have a dramatic impact on your personal emissions. Practical science is a product of honest conversation and is what delivers predictable outcomes for you in your world.

Earth Is Pre-Disposed to Friendliness Unless Pushed

If human society is to progress, Earth's climate simply has to be maintained in the stable and favourable zone it has occupied for the past 12,000 years. The climate and biophysical dynamics humanity would face outside of this narrow range would almost certainly put a very nasty end to the progressive civilization our very numerous and now well- armed populace has created.

We need to develop coherent strategies for our own lives, communities and nations. This will only be possible once we step out of our current rapacious consumer mode. At that point, we can ignore the self-serving advice from the growth-dependent parasitic overhead and start being kinder to ourselves and to the planet on which we live.

7

How Do We Get to Where We Need to Be?

Sustainable

Why Renewable Energy? Why Sustainability?

We have the choice of two paths going forward.

1. One is to fit the entire human enterprise into a package that can be sustained by our earth's natural systems.
2. The second is to believe that natural systems are but a short term stepping-stone to a future of galactic possibilities enabled by never-ending technological breakthroughs.

The first requires that we understand earth's systems and rapidly change our ways to prevent the collapse of the only biosphere we know of in the universe capable of supporting human life. This path does not shun technology, it depends heavily on increasing our efficiency and general technological prowess. It also means that we rapidly develop the technology and culture of sustainability.

The second path assumes that unless endless growth is maintained, economic and societal collapse is inevitable. Therefore, we must continually produce and consume more and eventually replace all natural systems with human engineered mechanisms. This will be made possible because human learning is unlimited. The expectation is that now we have crossed the threshold from ignorance into science, we will be able to innovate our way to any outcome we choose.

© The Author(s), under exclusive license to Springer Nature Switzerland AG 2022
J. E. Meyer, *The Post-Pandemic World*, https://doi.org/10.1007/978-3-030-91782-1_7

Value Systems

Confucianism rests upon the belief that human beings are fundamentally good, and teachable and improvable through personal and communal endeavor. It takes a village to raise a child. **Confucianism** focuses on the cultivation of virtue in a morally organized world.

Can a "morally organized world" be expanded to include the biosphere and the well-being of future generations? Perhaps this has always been the case but it is just something we've recently forgotten in our globalized and monetized world.

This assumes an endlessly expanding human population that will outgrow earth and expand across galaxies. People will live in huge man-made planetoids as we outgrow the surface areas of all the planets within our domain. Think of a traffic jam in space of the Borg cubes of Star Trek fame. Apparently Jeff Bezos adheres to an outlook along this line but he would be in the company of many commercial economists.

However, this book is for people who believe that we have to fit onto this earth rather than using the last of its resources to fire off 9000 space shuttle flights daily[1] to find destiny in space. Even if we had the fossil fuel reserves necessary to power 9000 large rocket launches per day for a century, it is unlikely the atmosphere or the ozone layer would be human friendly after even a few decades of this level of abuse.

Endless Time, Endless Possibilities

When problems are not pressing, good solutions can be arrived at over time with a minimum of cost and disruption. Perhaps even near ideal solutions can be delivered if the vision from the outset has been largely correct.

But we don't have an infinite amount of time and our dynamic world makes it impossible to predict what the ideal course might be. Our very late start on carbon reduction means that we have to choose direct and sustainable options. Since we are 30 years or so behind the curve, a slow winding down of bad practices via switching to less bad practices simply isn't an option. In the 2020s and beyond, all initiatives will have to be positive when they hit the ground.

There is no point improving the average gasoline fuel efficiency of the ground transportation fleet by 20% by 2030. That would be a good thing but

---

[1] 9000 space shuttle flights with 25 people/flight and with no supplies would allow the earth's population of 8 billion to be launched into space over a period of 97 years.

not remotely good enough. What has to happen is that all new vehicle production will have to be electric as soon as manufacturing capacity can be brought up to speed. If we make slow improvements to the fossil fuel efficiency of cars and trucks, the population of internal combustion engine (ICE) vehicles will slowly decline and (given an 18 year life span) linger on well past the middle of the century.

By 2050, the entire ground transportation network needs to be electrified possibly aside from special use vehicles. By 2050, most buildings need to be energy neutral and hopefully, energy positive. Recycling needs to be a very high fraction as opposed to the current very low fraction of waste and product cycles need to be several times longer.

In sum, we need be pointed in the right direction, act quickly and decisively and be tolerant of setbacks. We have to understand that we will be learning as we go. The less time we leave ourselves, the more assets will be stranded leading to greater economic and social disruption.

Real World, Limited Time, No Waiting for Silver Bullets

But in our current timeframe which calls for near complete decarbonization of our societies within the next 30 years, stumbles, mistakes, conflict and high costs are inevitable. In order to give ourselves as much leeway as possible we need to, as New Zealand PM Jacinda Ardern put it for her Covid-19 strategy, go at it "early and hard".

We use averages to determine how much of something we need but we don't live by averages, we live by being able to cover the extremes of consumption. The more extreme the climate of a particular region, the more necessary it is to take the extreme highs and lows into account.

For instance, the temperature range for Holguin, Cuba is 25 °C (highest 35 °C and lowest 10 °C), while the temperature range for Fort McMurray, Alberta is 92 °C (high 39 °C and low –53 °C). A day at the most at 10 °C is unlikely to kill many people or crops in Cuba but at –53 °C, a breakdown of machinery or shelter can prove fatal in mere hours. Lethally high temperatures have recently occurred in the western portion of North America as well as in northern Russia.

Although global warming has yet to have its Pearl Harbour moment when public opinion galvanizes in hours, the herd is becoming unsettled by the exceptional climatic events of the past few years.

Humans are much better at responding to disasters than distant emergencies. Unfortunately, we need to respond decisively during a climate emergency to prevent a climate disaster. Responding during or after the climate disaster will be too late. Climate rescue has to be now.

Can we do this? Yes. Rather than throwing our hands up and figuratively taking our masks off and driving to the Karaoke bar in our 4 ton personal transports, we are indeed capable of addressing these long term threats.

But we need to get over the greenwash and start to have adult conversations about what the job entails. Sustainability is hard work and conscious effort is required to build the necessary good habits. For instance, replacing plastic bags with tote bags is seen as a way to eliminate a great deal of plastic waste. It is in fact a good way to eliminate plastic waste but only if the tote bag life is in excess of 150 uses. So if you can use your totes for at least 3 years, you are indeed making a contribution[2] but it takes time and gains are small.

Working toward sustainability will not be like sinking a pipe into a virgin oil field and standing back to avoid being drenched by the gushing oil. Returns of this transition and downsizing will be thin and the reward will be retaining the core elements of what our society needs to endure and prosper.

We need to have the awareness that all things are connected and all physical problems have biophysical causes. We need to connect the dots and embrace reality no matter how ugly it may appear and what changes it demands. Do we really need to continue pulling the trigger to confirm that bullets are indeed going into our foot?

- The Mystery of the Missing Husband

    - Woman to detective – my husband is missing and I want you to find him
    - Detective to woman – what is your husband's name
    - Woman - Frank
    - Detective – how long has Frank been missing
    - Woman – ever since the funeral
    - Detective – whose funeral?
    - Woman – Frank's funeral
    - Pause
    - Woman – I want you to find him please.

The great years of the fossil fuel era are dead. We need to embrace a future without them.

---

[2] https://www.euronews.com/living/2020/01/22/is-your-reusable-tote-worse-for-the-environment-than-a-plastic-bag

Resilient

Extremes Happen

Problems rarely come one at a time and sometimes Black Swans arrive in flocks. The less well-informed governments are, the fewer biophysical processes they monitor and plan for and the more shocks their constituents will experience.

From CNN: "4 Million Texans without Power"[3]

- "That nightmarish supply-demand situation has sent electricity prices in energy-rich Texas to skyrocket more than 10,000% compared with before the unprecedented temperatures hit."

The Great Texas White-Out is an example of an extreme event we just aren't prepared for. Hurricanes are a well-known and expected event. It is impossible to predict when or where they will strike but no one is surprised when they do strike. The Great Texas White-Out delivered a multiplicity of cascading impacts from failing power grids to burst watermains and domestic water pipes to house fires and cancelled surgeries that resulted in the deaths of 210 people.

But as shocking as it was, this weather event did not do long-lasting damage to the grid or generation infrastructure. Contrast it to the Great Ice Storm of 1998 that crumpled 1000 hydro transmission towers in Quebec and snapped 35,000 wooden hydro poles. Four million people lost power and millions of trees were killed or damaged. Three weeks after the event, 150,000 people had yet to have their electricity restored.

The Texas grid should have been winterized but it wasn't. That was a government decision. But no matter how well prepared a system is, there are some events to which it will always be vulnerable (Figs. 7.1 and 7.2).

Hailstones of this size will shred[4] and lay waste to a wide variety of infrastructure from roofs of all sorts to siding to automobile windows and body panels and from crops to telecommunications antennae. And, of course, the larger the hailstone, the faster it falls multiplying its destructive potential.[5]

---

[3] https://www.cnn.com/2021/02/16/business/texas-power-energy-nightmare/index.html

[4] https://www.abc.net.au/news/2020-01-28/solar-profits-threatened-by-nem-rules-killing-investment/11903706

[5] https://www.nssl.noaa.gov/education/svrwx101/hail/

Fig. 7.1 Renewable doesn't mean risk free

Fig. 7.2 Extreme events are capable of wreaking havoc with every part our infrastructure

The average area covered by a hail storm can range from 1 or 2 sq. km to hundreds of sq. km.

Something Truly Unusual

Most people think of downbursts or straight line winds as powerful but localized events. However, an event called a derecho occurred in the US Midwest on August 10, 2020 and covered a huge area measuring some 1200 km long by 300 km wide with winds of up to 200 km/h, 5 cm hail and 25 tornadoes. Stretching from eastern Nebraska to western Ohio, it caused $11 billion in damage, killed four people and destroyed 340,000 hectares, or 1300 square miles, of crops in Iowa alone (Fig. 7.3).

The most secure people realize they can never be prepared for everything.

What Texas Didn't Have: Resilience

At some point, resilience is impossible but, going forward, we need to understand that we are coming from an era of exceptional climate stability. We have also enjoyed the most stable energy supply in human history. We are now entering an era in which climate will be more unstable than in any period of human civilization. Additionally, we will be transitioning to energy sources which are far more variable and harder to store than the fossil fuels we've come to depend on over the past 400 years.

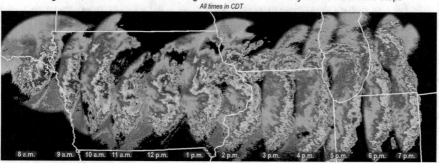

Fig. 7.3   There are extreme events and then there is the unique Derecho

The reasons for Texas' lack of resilience in the face of this winter weather event comes down to a number of missing links which would have mitigated the severity of the crisis:.

- Interconnected grid
- District heating
- Geothermal storage
- Winterized wind, natural gas and nuclear systems
- Winterized public water systems

Deep resilience is needed because once unexpected things begin to happen, problems cascade. Austin Texas Mayor Steve Adler; "It feels like it's just one thing after another after another."

"We were just not prepared for these cold temperatures. We have a deregulated power system in the state, and it doesn't work." Lowest-cost based, just-in-time markets work in stable times but are not structured to deal with extreme events.

Panic (and Politics) Set In

Fox News'- Tucker Carlson (February 16, 2021) blamed the entire problem on wind turbines freezing. "The windmills failed like the silly fashion accessories they are." Wind generators work reliably all over the world in places like Denmark and the North Sea but the ones that froze up in Texas were not winterized. Neither were the natural gas, coal or nuclear plants which had to either reduce output or shut down.

And during the same cold snap, wind turbines in Alberta, which have been designed for extreme temperatures, worked as well as ever even though it was another 10–25 °C colder than it was in Texas.

Carlson continued stating that "Green energy means a less reliable electrical grid" while failing to note the isolated Texas grid is unable to either import or export electricity. Green energy must also include flexible grids, an array of sources and storage infrastructure to build resilience to deal with short term events.

But when Carlson noted that "People who support wind farms live very far from where the windfarms are going to be located. People who live near where the windfarms are going to be located have a different view", he spoke the truth.

There are disconnects between the people who advocate certain solutions and the people who have to implement and live with them. Without a common core of facts and understanding, every problem will be taken as a reason to give up and every setback will be seen as a disaster.

Socially Cohesive

Who Has a Chair at the Table: Leading Wealth Creation Sectors

In many western societies the level of equality is declining. Large numbers of high paying jobs have been exported to cheap labour countries. Combined with the increases in housing and energy costs, millions of previously prosperous individuals find themselves hard pressed and in financial decline. Failure to take the precariousness of these very large groups into account will result in social unrest which could grow into turmoil.

The Yellow Vest protests in France, the lockdown riots in the Netherlands and the Capitol Hill riots in the USA were sparked by different events. However, a common underlying factor is the general decline in the welfare and quality of life of large sections of the population.

Transitioning to renewable energy and moving towards sustainability will re-shuffle the economic and lifestyle cards of large numbers of people. If the changes are implemented in a way that produces a few winners and a lot of losers, the level of discontent may well reach destabilizing levels.

As coal production has wound down in Alberta and Germany, both governments have implemented extensive re-training and re-location programs. It is necessary to support workers and families through a transition to another field of work by assuring soft landings and smooth takeoffs.

To be successful, transitions have to involve all of the productive, real wealth creation sectors of the economy:

- Labour
- Science
- Small and big business
- Farming
- Energy & Resources
- Transport
- Environment

- Education
- Health

These wealth driving sectors need to sit at the table while supporting sectors of finance and construction would contribute as the evolving market requires. Media corporations need to deliver the national conversation intact and not filtered and mutilated to suit their own financial agendas. Developers and finance cannot be allowed to determine national goals which create an ideal market environment for themselves any more than tobacco companies should be allowed to write health legislation or the arms industry be allowed to determine foreign policy.

Losing the "Me" Attitude

Consumers have been taught to believe that their immediate needs and wants are always top priority. Rights to all manner of goods and services are frequently seen as sacrosanct while responsibilities are rarely mentioned. In many cases, responsibility and long term thinking are bad for business but social well-being needs to make it back into the public psyche.

Consumers can transform into citizen heroes within minutes of a disaster striking but can they maintain this ethic over the long haul or will "transition and responsibility fatigue" set in as Covid fatigue did for some groups. Fear of not having the brightest smile, fastest car, biggest house or most toys, is a tough attitude to change but the pandemic has shown us that these are not the things that really matter.

Post Covid-19, do people really have to still be told what fashion accessories they need to live their lives? Many individuals have learned how to distinguish between the quality of life and the quantity of material goods. Now governments need to identify true progress as distinct from ever higher levels of commercial activity.

There are very solid reasons for not believing what comes out of the mainstream media where economists are put on the same level as scientists and commercial economics is given vastly more play. But commercial economics is not science and the constant blaring of its misguided and transparently self-serving messages has eroded confidence in the entire corporate media information flow.

Instead of being used to describe human welfare and social and environmental balance, economics has become a platform controlled by narrow interests to increase their power and wealth at the expense of every other interest around them. Leadership has to come from broad-based interests.

Our New Toolkit, Our New Metrics

Measuring what is important is critical and our understanding of what is important has shifted dramatically over the past 50 years to include the health of Earth's many biosystems. Since we now process a large proportion of the planets energy and biomass, we need to pay as much attention to the waste we produce as to the products that created the waste (Fig. 7.4).

We are starting to be able to monitor recycling effectiveness which has huge implications for the issues of resource scarcity[6] and environmental decline. The days of industries taking the money and running are coming to an end as lifecycle standards are implemented for an increasing range of extraction and manufacturing processes.

Consequently we are learning where to put our scientific investment efforts and these are turning out advances in many areas such as plastics made for

Fig. 7.4   Better recycling rates depend on citizen involvement

---

[6] Chris Clugston's "Blip" examines resource scarcity on a mineral by mineral basis.

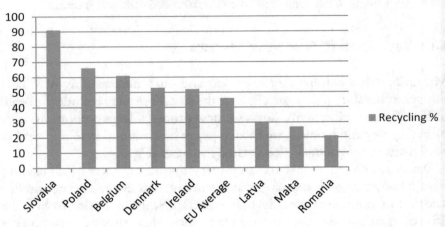

Fig. 7.5 Different countries, different recycling rates, many with lots of room for improvement. Europa.eu Eurostat. (https://www.epa.gov/sites/production/files/2016-11/documents/2014_smmfactsheet_508.pdf)

ease of recycling. If the future is recycling then once is not enough.[7] Will this new process actually recycle plastic hundreds of times without losing the quality of the material? That is not a perfect solution but it is progress (Fig. 7.5).

In 2014, the U.S. Environmental Protection Agency (EPA) reported the rate of lead battery recovery was almost 99%. The highest recycling rates for other well-known products were: newspapers (63%), aluminum cans (55.1%), tires (40.5%), glass containers (32.5%) and polyethylene terephthalate (PET) bottles (32.2%).

The 120 years of experience we have in dealing with lead acid batteries has enabled us to learn how to recycle them efficiently providing cause for optimism regarding our potential to recycle all material at very high levels.

Asphalt shingles are a significant source of landfill waste but steel roofing has a much longer useful life and it is easily recyclable. Steel roofed buildings are also far more resistant to damage from forest fires, which are an increasing threat in areas of Australia, North America and Europe where many communities are built right into forest areas. The trends to both improved recycling rates and the switch to more long lasting and recyclable materials are clearly on.[8]

---

[7] https://www.euronews.com/living/2020/02/04/breakthrough-means-plastic-can-be-recycled-hundreds-of-times

[8] At the moment, around 45% of plastic waste is recycled annually in the UK and is on the increase. https://www.forbes.com/sites/niallmccarthy/2016/03/04/the-countries-winning-the-recycling-

When will the move of manufacturing back to "developed" countries begin in earnest? The environmental and labour standards in many cheap labour countries can no longer be ignored. Consumers in the "rich" countries must take into account the environmental impacts of the goods they are consuming, not just the impacts of the goods they produce on their own soil.[9]

The pandemic has brought out calls for the renewal of the industrial base in countries which have lost theirs over the past 4 decades. The inability to supply critical medical gear and vaccines has shown the dangers narrow based economies pose to public safety. Add in higher environmental and recycling standards and the appeal of self-sufficiency is compelling especially when the lens is widened to include food.[10]

Institutional Corruption in Various Countries: Ability to Make Progressive Decisions

Different groups make their living from different activities and some groups acquire a great deal of influence over policy making which they use to make market conditions more favourable for their particular business model.

Policies favouring one industry may well hurt another industry. "Dutch disease" is a term used to describe one export industry booming and increasing the value of its country's currency which unfortunately puts another domestic industry at a disadvantage.

Canada's "petro currency" is a good example of this effect. Exports of oil amounting to around 30% of Canada's exports pushed up the value of the Canadian dollar by as much as 25% and consequently put Canadian manufacturers at a 25% disadvantage in foreign markets. Hundreds of thousands of manufacturing, tourism, agricultural and forestry jobs were lost across the country.

Norway managed its "oil bonanza" on a national basis so the infusion of petro-money didn't hurt any domestic industries. He who pays the piper naturally has a strong desire to call the tune.

Similarly, politicians in and around Canada's urban centres are dependent on donations from developers leading to a "growth mandate" in most

---

race-infographic/#2dc69d1a2b3d

  https://www.weforum.org/agenda/2017/12/germany-recycles-more-than-any-other-country/

  https://www.lovemoney.com/gallerylist/89902/countries-with-the-best-and-worst-recycling-rates

[9] https://www.abc.net.au/news/2020-07-23/coronavirus-pandemic-leads-to-australian-manufacturing-revival/12481568

[10] https://en.wikipedia.org/wiki/List_of_countries_by_food_self-sufficiency_rate

municipal governments. These same conditions exist in Australia where coal takes the role of the country's other large policy influencer.

The most powerful businesses seek to have their run at the top extended for as long as possible and are not likely to put public interest at the top of their priorities. In a broadly based economy, a discussion of well-being is much easier to hold than in a country dependent or controlled by one or two dominant industries. It is vital to identify who holds sway over public discussion and public policy formation.

Self-Aware

Different Latitude, Different Destinations and Means of Getting There

Understanding the strengths and weaknesses of the area you live in will greatly assist in your ability to be proactive. The American Southwest gets highest solar PV output during the summer when they need it most while in northern locations, solar PV output is lowest in winter when it is most needed.

Communities need pro-active leaders. Will COVID-19 kick-start a green renaissance – or stunt it?[11] Will desperation to make up lost economic round throw the environment to the wind or has the pandemic experience instilled a more holistic perspective?[12]

If leaders even recognize they are on a treadmill, do they know how to get off it? How many times do the USA, Canada and Australia need to repeat the mining town collapse experience before leaders wake up to the fact that what works for small towns also works for nations. Mining and resource depletion is not a basis for a stable economy. Nor is asset inflation and money printing.

Follow the Money

The power behind the throne knows where their money and power come from. Anyone seeking to change social priorities needs to understand the origins and flows of money and power both locally and nationally. Those seeking

---

[11] https://www.theweathernetwork.com/ca/news/article/covid-19-coronavirus-could-kick-start-a-green-renaissance-or-stunt-it

[12] https://www.euronews.com/2020/03/25/coronavirus-leads-to-a-sustainable-society-better-equipped-to-tackle-future-problems-view

Fig. 7.6 The choice is ether to die with a stack of bitcoins in your account or live well in a healthy community

to promote systemic change need to identify the interests of the people currently driving decisions and put them into public view (Fig. 7.6).

The Simple Approach Isn't So Simple

How do we make sure we get to where we need to be in time?

1. Start doing what is good.
2. Stop doing what is bad.

One can picture a business-as-usual politician standing at a podium and suggesting to his dear friends in the audience that now is the time for action. Staring intently over the audience, he pounds his fist on the lectern and demands that we first stop doing bad things and then start doing good things.

The opposing party business-as-usual leader responds with incredulity and declares this to be the policy of defeatism. He gently posits that an enlightened approach would entail starting doing what is good and then stopping to do what is bad.

The debate spirals down into the polarization of opposing political parties with the do-good first and stop-bad second camp locked in a bitter power struggle against the stop-bad first and do-good second camps in a battle lasting for decades. Nothing happens and the winner is the business-as-usual interests which fund both parties.

Unlike with the Covid-19 crisis there are extremely powerful interests who both want and need to make sure no effective action is taken on climate change and the host of other biophysical problems this planet faces. A fundamental shift away from society's singular focus on the growth of the commercial market is critical for human welfare and environmental health. However, a change away from simple growth will be a crippling blow to the power of the current developer, debt, finance, cheap labour and media corporation elites which now control public debate and public policy.

The farcical debate portrayed above may seem completely fanciful but unless real hard numbers are being used to illuminate our problems, a good many national conversations are every bit as dysfunctional. The number one priority of interests working to derail meaningful policy making is to prevent science from informing the debate and displacing population and consumption growth as the top national priority.

Their control of the national policy apparatus plays out daily on the mainstream media platforms as political theatre. This process features professions of alarm and concern, bold new photo-op announcements and ribbon-cutting events which are as ineffectual as they are rehearsed. A lack of science, clear physical and social goals and solid numbers should start red lights flashing just as would a mining stock promoter's pitch on a company without any financial statements or assay results.

Your Reality Versus the Reality of the Growth Dependent Sector

Politicians like growth. They like to point to shiny new buildings, building cranes, sprawling subdivisions and broad highways and larger budgets. To them this represents a healthy community – one that is vibrant with new projects and ever more people. They speak in terms of "Barrie is booming" or "We're growing faster than New York City" etc.

But cities and towns are not sentient beings and they have no rights or feelings. Cities, towns and countries are social structures designed to serve the interests of their residents, not the other way around. If politicians have lost sight of the fact that their obligation is to improve the lives of their constituents, then they have to either be reminded of their responsibilities or replaced.

For all too many politicians, growth is mission accomplished. They look no further than the political donations which flow in from developers and speculators along with the favourable coverage from growth dependent media corporations.

Citizen life satisfaction, affordable housing, clean and secure streets, better jobs, modern infrastructure and ready access to nature and healthy walking trails and parks are what mean the most to people. Rapid growth, housing inflation,[13] more traffic, cheaper labour, and denser, more crowded urban cores are the factors which put cash in the pockets of the interests who promote "bigger-is-always-better".

If the population of the town/city/country stops growing and stabilizes along with housing costs, speculators and subdivision developers will be out of business while media corporations will see a large chunk of their advertising revenues disappear. If the government then devotes all of the investment money it was spending on expanding infrastructure to improving infrastructure, then residents can enjoy an improved quality of life and a more cohesive community.

These two sets of interests are diametrically opposed and politicians and the media should be called out whenever they conflate growth with progress or use the town's size or growth rate as a proxy for the well-being of the people who live in it.

Real goods and service producers like the growth of sales as well but they don't need it to survive. A farmer doesn't need to produce 3% more corn every year, nor a Doctor examine and extra 40 patients or a manufacturer punch out 10% more widgets annually. A stable economy holds no fear for real wealth producers but it spells death for growth dependent interests.

How Will Widespread Adoption of Renewable Energy Affect You Personally?

The problem is bigger than changing lightbulbs and unplugging electric toothbrushes. Are you ready to make significant big and small changes to reduce your footprint? Welcome to life in continuous transition.

Expert Quote: "Fuel switching is NOT an easy thing to take on, particularly in a world of changing supplies, industries and technologies. It's funny because just a couple years ago LED's were the low hanging fruit... People thought they were saving the world by putting them in their homes... that's how disconnected from reality they were. We need to steer the conversation to the major emissions sources (waste, transportation and heating)."

---

[13] "housing appreciation" or "the wealth effect" or whichever term promoters use to sell debt and unaffordable housing.

Given the wealth of carbon and energy calculators available on the internet, building a budget is relatively straight forward. You can compare your performance with the national average and with averages of those in other nations.

Footprint calculators[14] are available for most aspects of our lives and make great family and school projects. How much energy is your household, your community and nation using now? What is the carbon footprint of your child's school? How many tonnes of carbon do various activities produce?

- Home
- Car
- Material goods
- Food
- Travel

By developing a feel for the energy flows and material consumption in your life, you will be able to make solid choices on how to invest your time and money in reducing your impact on the planet. Developing the tools our children need to survive and prosper with complete resource, energy and environmental lifecycles is the most valuable legacy we can pass down.

Learning

Face the Future

Science has brought awareness of our environmental problems into the public forum but science is being substantially locked out of discussions of solutions or pathways to the end goal of sustainability in many countries. One look at the range of government "green initiatives" in countries from Canada to the European Union to the USA shows a lot of greenwash but few concrete steps of the necessary scale to build the renewable energy infrastructure to get us to the goal of sustainability.

We have to position ourselves and our communities for success on the broad array of challenges we face. The seemingly disparate issues of energy availability, low carbon emissions, species support, recycling, social cohesion and many others will suddenly begin to enmesh and collide if we fail on any one of the major facets of the transition to sustainability.

---

[14] https://www.gbbr.ca/carbon-calculator/

The next chapters are devoted to how we manage our own lives to step onto the path of sustainability and how we might influence other individuals and governments to make a common push for security and resilience in our very dynamic and shifting biophysical world.

Setting the Table with Models

Perspective and properly weighted measures can best be enabled by mathematical models. These models are based on what we have learned over the past decades and centuries. Their projection mechanism is calibrated using all of the historical data that can be assimilated.

The framework of assumptions these models provide is open for all to examine and once hard data rolls in, real-time modifications can be made as the model acquires the knowledge from "the living labs" we will be living in.

Scientific guidance is critical because we are dealing with biophysical issues which are subject to physical laws. King Canute followers and those who feel they can dictate to nature or appeal to a higher power for special treatment need not apply for leadership positions. The nations which learn the fastest, act on the best information and fix their mistakes diligently will have the easiest time in the transition to sustainability.

Our Lives in Nature, Our Responsibilities to the Future

Spanish saying: "God forgives all, people forgive some, nature forgives nothing." God may have designed the earth but Mother Nature runs it and we need to be cognisant of the rules by which we ultimately live. That does not mean abiding by these rules needs to be odious.

Recent studies conducted during the pandemic period have shown that people are happier and healthier if they have substantial exposure to the natural world. Most of us knew that all along.

Taoist thought focuses on genuineness, longevity, health, vitality, equilibrium and transformation. These principles are repeatedly stressed in all manner of self-help and life affirmation exercises. We will all feel better if we can leave the world a better place which is the equivalent of the Japanese ethic of "son gets a better rice field".

Ecological civilization is now written into China's constitution and Bhutan has implemented an Index of Happiness as the main metric for government policy making. Eventually multi-generational thinking will return to our consumer society and displace its short-term commercial outlook. But first we have to recognize that we need to stop heading in the wrong direction.

A saying which should apply to the Alberta oil sands and to our consumer society sums it up:

> *"If you are in a hole that is getting deeper, stop digging."*
> *Will Rogers*
> *(also repeated by Winston Churchill)*

Leadership

Leadership is the vital ingredient that combines all of the above factors and moulds them into a coherent effort to deal with the challenges ahead. Able leaders will "get on it early" by investigating and developing the information and expertise to make the necessary decisions.

And when they have enough information to take decisive action, they "get on it hard", learn as they go and don't let up.

Leadership is needed at all levels of society, from independent individuals to community and national leaders.

China has leadership. Over the past several decades it has been able to set clear goals enact successful strategies to achieve them. If development and national policy execution were Olympic events, China would be wiping the floor with the western democracies.

From Covid-19 response to hitting its social and economic targets, China's record is unmatched for a generation. We in the west like to think of ourselves as advanced in terms of democratic and individual rights but China is winning influence around the world with its aggressive investment and foreign policy initiatives.

In contrast, the west has given away its industrial might and technological leadership along with the welfare of large chunks of its citizenry in the name of globalism. Social cohesion is unravelling while its elites seem to profit ever more from the decline of lower income groups. Democracy's image is now tarnished.

China has bought huge new influence and control of vital resources around the world with the money western democracies gave it to make cheaper versions of goods they were capable of making themselves. At the same time, it has delivered vastly higher living standards and a much higher quality of life to the great majority of its people.

Unsurprisingly, given its history over the past 200 years, social order is a top priority for the Chinese government. They have implemented extreme measures to assure their revolution stays on track and that compliance with Communist Party directives is complete. The sometimes draconian application of state priorities exacts a great cost from those who do not completely share the states view of the world.

However, the success China has experienced and the influence it has gained stands in stark contrast to the confusion and upheaval western countries are experiencing and for which there seems to be no end in sight.

The west has embraced a laissez-faire system in which there are no standards or goals, just the deal-of-the-day ethics. China does not play by the same rules but is using the west's hands-off approach to put its hands on the instruments of power that have been abandoned by globalist governments.

In authoritarian regimes, the national conversation is one way. In the democratic model, the first responsibility of leaders is to assure the national conversation is open and transparent. Otherwise "democracy" simply devolves into a chaotic, low functioning authoritarian system.

But in democracies, in the absence of political leadership, individuals are free to exercise their own. Clearly individuals need to manage their own affairs and drive change in their communities in order to make the transition to a sustainable society work. The transition to a resilient future does indeed begin at home.

8

Your Home and Lifestyle

Compared to human experience throughout history, in 2022 we are living in larger, more comfortable homes, eating much more and higher quality (if we so desire) food with health care and education infrastructure better than our ancestors could have imagined.

One of the largest of these advances is the increase in mobility we enjoy. Using personal vehicles, trains and aircraft, it is now practical to experience the thrill of movement over vast distances in near-perfect safety. The most ubiquitous of these modes of transport is the family car which has transformed our cities and the way we both work and live.

The sophistication and comfort of our homes has also increased immensely and they are safer and healthier than they ever have been while using less energy.

Individuals drive change and it is critical that people recognize both the extent of their power over their own consumption levels and the limits of governments to adapt to any situation and fulfill all the expectations of its citizens. People need to take responsibility for their own ecological footprint.

Whatever you decide to do, please keep in mind there are eight billion of you. The accelerating pace of unexpected events may leave those assuming governments have effective strategies to deal with all challenges. Were they ready for an easily identified pandemic?

J. E. Meyer, *The Post-Pandemic World*, https://doi.org/10.1007/978-3-030-91782-1_8

Major Components of Your Lifestyle

In resilience preparation, as in dieting, the gradual approach works better than taking extreme measures. Some North American Plains Indian tribes practiced the Sun Dance ritual during which braves were suspended by hooks embedded through their chest flesh. Only the most committed individuals subjected themselves to this practice. 200 years later, the sustainability equivalent of the Sun Dance ritual, namely moving into a dugout cabin without running water or electricity, is a level of sacrifice beyond the pale for all but the most ardent green advocate.

Fortunately no one needs to go to those extremes given the advanced and proven products and designs now readily available. Your home is the place where you spend the most time and have invested the most money. It is also where the majority of your energy budget is spent.

Transportation, material goods and food consumption make up the rest of your ecological footprint but tweaking where and how you live requires the highest level of involvement and can produce the highest level of environmental benefits. But for a fast start, replacing a conventional gas vehicle with an EV can't be beaten.

The 1, 2, 3 and 4 easiest ways to reduce your long term costs and immediate ghg emissions and footprint as well as improve your own resiliency are:

1. **EV**
2. **Geothermal heat pump**
3. **Solar PV and hot water**
4. **Storage – heat and electricity**

What Is Meant by "Geothermal"?

There are three different types of geothermal systems. They all involve piping systems buried in the ground to either extract or store heat.

- The most common is a **geothermal heat pump** with its piping installed in the ground which has an average temperature between 8 °C and 25 °C. These systems transfer heat from the ground into the building.
- **Geothermal storage** uses the same hardware as the basic system above but is capable of pumping heat into the ground during summer for storage to be used months later during winter. The ground temperature can then range from 8 °C to 60 °C.
- **Geothermal power or energy systems** require very high heat to drive steam turbines which generate electricity. Typically a ground temperature in excess

(continued)

of 180 °C is needed and there are relatively few geographic locations which can produce enough reliable heat to be practical. Pipes must be sunk 2000–5000 m deep to reach a 180 °C rock stratum.

 - Conversely, suitable ground conditions for geothermal heat pumps and storage systems, are abundant and widespread.

Heat energy is very different from electrical energy and is best suited to different tasks. Heat is cheaper and easier to both generate and store than electrical energy. It will play a prominent role in our transition to renewable energy.

Strategy

How would you have organized your life if you had had 2 years of forewarning of the Covid-19 pandemic? Or similarly for:

- The wildfires in the Western North America, Australia and Europe
- The repeated hurricane strikes in SW Louisiana?
- The floods in many areas of the world?

Well, you know climate change is coming and you know that the transition to renewable energy will be complete by 2100 and largely in place by 2050 so what does that mean for your long term planning?

If government commitments to go almost totally carbon free or carbon neutral (two very different things) by 2050 are actually realized, the way you live and work will undergo a substantial transformation. These changes have to start now. There are several things you can count on if the transition is successful.

1. Renewable electricity will be cheaper than fossil fuels. Whether this is achieved by pure market forces or pure government mandate is irrelevant.
2. Energy will be more expensive than it is now.
3. Your community and probably your home will be an integral and intelligent part of complex energy flows.
4. Recycling rates will be much higher.
5. Your home and business will only have one high grade energy connection: a wire to the electrical grid. Natural gas hookups will be phased out.
6. Your private vehicle will be electric and might possibly have pedals.

If an orderly and positive transition to renewable energy does not take place by 2050, we can count on increasing social chaos and debilitating weather events dramatically reducing the level of civilization to far below that which we currently enjoy.

In the end, individuals and families need to consider what it will take for them to ensure energy and food security, as well as a healthy community and lifestyle. Resilience to buffer the worst effects of extreme weather events and energy interruptions will also be key. However, few people have the unlimited resources needed to be able to glide into a sustainable future.

A real-world strategy for most people will include a close look at:

- Conserve - reduce your energy demand load
- Generate as much of your own energy as possible
- Store as much energy as possible
- Don't be caught with stranded assets
- Resilience is a cost you may always have a hard time justifying but, when you need it, is a benefit you can never put a price on.

Strategize about where and how you live, what you drive, and what kinds of energy you consume and how much you could produce. Some things you can change easily and some things will prove difficult.

Does it make sense to move to a different community with lower energy demand and higher levels of renewable energy production potential? Reducing your demand by conservation measures takes a great deal of pressure off all the downstream decisions and investments you will make.

If you currently have an aging oil furnace needing replacement, will you buy another oil furnace or invest in either a ground-source or air-to-air heat pump? When looking at future trends, this is an easy decision to make.

Picking a sustainable city or region[1] is the most fundamental and difficult choice a person can make. It has profound impacts on every aspect of their life well beyond the technical issues of energy and resilience. But you really do want to be in a healthy community with good access to nearby food production.

Saving for retirement? Invest to reduce your cost of living. Know what the costs and trade-offs are of investing in generation vs conservation. And to minimize financial losses on "stranded assets", make sure you invest in renewable technology when it is still your choice to make. If you wait until you are forced to buy green compliant cars, heating systems etc., you may well find

---

[1] America's Most Sustainable Cities and Regions, John Day and Charles Hall, https://www.springer.com/gp/book/9781493932429

that your old fossil fuel or inefficient products have to be written off even if they might still have a great deal of life left in them.

Don't buy an oil furnace with a 20 year life expectancy just 2 years before the use of oil for home heating becomes either illegal or punitively expensive due to the implementation of higher carbon taxes or stricter insurance regulations. When it comes to efficiency, conservation and renewable energy, the wind is blowing only one way.

Disentangle your real needs from what the commercial marketplace tells you your needs are. As the title of an article[2] by Jil McIntosh on the Driving website asked:

"Okay, but do you really need a truck? Like, really?"

Or a 400 hp. racing sedan used to commute in heavy traffic? Or used solely on public roads. Or an all-terrain capable SUV which will never be taken off pavement or a vehicle with a capability you might only use every 2 years for which you can easily substitute a rental or delivery service.

"And just like Buddy, you'll soon find out your *need* was really just a *want.*"

For instance, here is a high tech energy toolkit for low energy living during cold weather:

• Socks, slippers, scarves, sweaters, woollies and hats, extra blankets, plastic window film, storm doors, window and door caulking.

When heat becomes a problem the toolkit flips over to:

• Shorts, t-shirts, sandals, fans, open windows, porches, cellars and shade.

Clearly there are practical limitations to the above but we have to change from just throwing energy at a problem to minimize it.

---

[2] If any of the following criteria apply to you, you probably don't need a pickup by Jil McIntosh.

https://driving.ca/ram/features/feature-story/okay-but-do-you-really-need-a-truck-like-really?utm_campaign=on_network_boosting&utm_medium=native_ads&utm_source=drivingcontentads&utm_content=Drivingflex&mvt=i&mvn=83464705e8e54c1ca5ea76cbc105a3a4&mvp=NA-NPFP20-11238683&mvl=nationalpost.com%20-%20index

Getting Your Bearings

How different are your needs and consumption patterns from the average and how can energy budgets be broken down into manageable segments?

Below are two graphs showing the pattern of energy consumption in American homes. American data is used because it covers disparate geographic regions which serve to illustrate the effects of location on consumption patterns (Fig. 8.1).

The above graph shows energy use per household declining over time with levels of energy consumption being very different for different regions. It is safe to assume that comfort and health generally improved over that 35 year timespan yet energy consumption declined by almost one third due to higher standards, technological advances and increased investment.

Was it worthwhile? Picture your household energy bills going up by 50% as they would have without those advances in conservation (Fig. 8.2).

Different types of residences have different mixes of energy consumption. A stand-alone house, with its four exterior walls, windows all around and a roof will use a higher percentage of its energy consumption for heating compared to an apartment in a large building.

Residents and owners of these properties face unique investment options. An apartment renter can't invest in solar panels on the roof or a geothermal heating system. Would they even be allowed to install an external roll-down

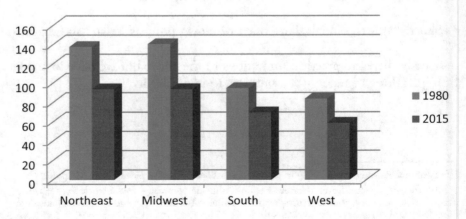

US Household Energy Consumption
(millions btu)

Fig. 8.1 Region residential energy consumption patterns

Energy Use In Different Types of Homes

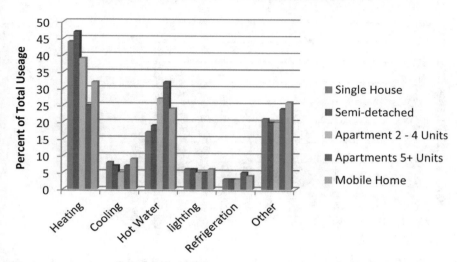

Fig. 8.2   Energy consumption by type of building

window shutter? Options exist but why would they invest in anything which was likely to last longer than their stay in that particular apartment?

The residential energy mix over the past 70 years has seen a shift to much lower carbon emission sources. Coal has almost disappeared from residential use and oil (petroleum) is in sharp decline, both largely replaced by natural gas. Meanwhile, electricity use has increased dramatically. As the future rolls onward, electricity's share will progressively displace even natural gas until (almost) all energy consumption is electricity. At the same time, an increasing proportion of electricity will come from renewable sources.

Residential Energy Use by Country (Fig. 8.3)

It is a fact borne of simple necessity that buildings in northern countries require more energy than ones in more moderate or hot climates. Sweden, for instance, has very high per capita energy consumption but, given its available supplies of hydro-electric power and it's very advanced geothermal heating infrastructure, its level of carbon emissions is comparatively low. This is something to bear in mind as we strive to both reduce our energy consumption and eliminate carbon emissions.

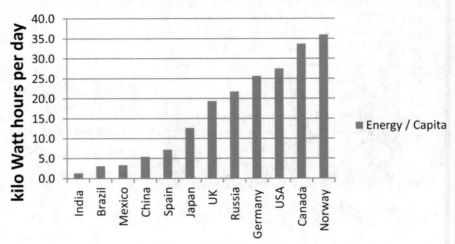

Residential Daily Energy Budget

Fig. 8.3 The more extreme the climate regime, the higher the home energy consumption

Your Home

The archetypical home in the western world is a single family dwelling powered by both electricity and natural gas and it has a fairly large internal combustion (ICE) vehicle sitting in the driveway. Over the years, the traditional family home has transitioned through a series of energy systems.

Initially, heat came from woodstoves and light came from oil lamps. These gradually gave way to the very similar coal stoves and revolutionary electrical lights. Electricity was a new energy "source"[3] which involved the transfer of mechanical energy from a distant source; say a waterfall, through a copper cable to the point of use. Electricity could be used to produce light, heat or mechanical energy.

Electricity began to displace combustion systems in the home when it began first producing light. It then went on to produce heat for cooking, mechanical action for washing machines and tools and low level power for communication such as telephones. It could also be used for space heating but it was expensive compared to the wood and fossil fuel alternatives.

---

[3] Electricity is a medium of transmission of energy, it is not an energy generator.

As fossil fuel development continued, first oil replaced coal and then natural gas replaced both coal and oil. It was cheaper and easier to use and distribute, burned cleaner with lower maintenance requirements and could also be used to heat water and cook food.

All through this process, the energy distribution system became more efficient evolving from wood carts and hand-bombing, to coal carts with sacks and shovelfuls of much more dense material. Then tanks of oil were transferred by hose and finally pipelines delivered natural gas which required no direct human involvement. Pipelines were fairly large and had to be put underground but electrical cables are comparatively tiny and very easy to "string" on poles and buildings. Welcome to "the Grid".

But through all of the steps in the maturation of energy systems, homes never produced their own power. To a greater or smaller degree, that is going to change in the near future because houses are constantly bombarded by energy in the form of solar radiation. They are also sitting atop an immense energy battery, namely the Planet Earth.

The potential of the sun and the storage capacity of the earth will soon assume prominent places in the communal grid. This heat grid will connect a wide variety of energy sources to a wide variety of consumers with very complex flows to and from every point on the grid.

In the end, housing and buildings in general will become close to energy neutral and may well become energy positive. Your home and its relationship with "the grid" is about to become much more complex, more efficient, and far more resilient (Fig. 8.4).

Your House Now

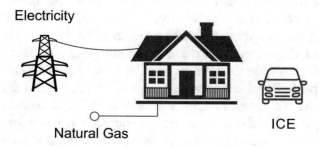

Electricity

Natural Gas

ICE

Fig. 8.4 Conventional energy grid of the twentieth century. Effectively Grid 101

Retrofitting

Even if the energy profile of your current abode isn't green and efficient and is in dire need of improvement, neither moving nor building new might be possibilities for a host of reasons. In that case, upgrading and retrofitting are the only options and they can be very good ones.

Retrofitting can range from next to no expense or time investment to substantial outlays spread over years. In a house in a northern climate with old and/or large windows or patio doors, an expenditure of under $100 and a few hours is all it takes to put up heat shrinkable plastic film on those windows and cut heating bills substantially. Comfort levels would also increased although some people might find the look unappealing.

Add another $100 for window and door caulking and sealing and results improve further. Many northern homes have only single exterior doors. These are big energy weak points. The installation of storm doors will cut losses from those entryways by 50% or more. Ditto storm windows although these have fallen out of favour due to their upkeep and storage when not in use. Also, their installation and removal probably requires the use of a ladder, a leading source of household accidents.

Summer time or southern climates reverses the issue and heat in the form of solar radiation, should be blocked from entering the house. External roll-down shutters or awnings are very effective here. Internal blinds will help but even though they block the sun, they are inside the house and therefore so is the sun's heat when it strikes them.

North-facing windows may be thought to be immune from solar loading but a considerable amount of heat does come through these windows when the sun is shining strongly on the ground or snow near them. An insolation meter is useful here to quantify the energy coming in the window but a person can fairly readily feel it on their face and hands.

Insulating the ceiling with either fibreglass, rockwool batts or cellulose might be the next cheapest upgrade to stem heat loss. In terms of energy saving for the dollar, the above upgrades provide by far the best results. Do-it-yourselfers can dramatically cut the energy consumption of their houses with those upgrades with less than 2 days work and for under $2000.

Hiring a professional contractor to improve the ceiling insulation will be more expensive but it will be safer and most likely yield better results. And certainly injecting insulation into the exterior walls is not a job for anyone other than a professional or the most experienced of home owners.

Insulating the foundation or adding underfloor insulation again may best be done by professionals unless the keen owner has a well thought out strategy to avoid trapping moisture and unintentionally promoting mold and rot.

The strategy is:

- Reduce heat loss
- Reduce heat loading
- Absorb energy when you need it and reject it when you don't
- Produce energy when you can
- Invest for the future
- Energy options and costs will be different in the future

It may not be possible to do everything all at once but the process can be staged by adding the upgrades as you can afford them to complement each other.

Next on the list is making the heating and energy system more efficient. Replacing furnaces and hot water tanks with a natural gas boiler for both space heating and domestic hot water is a large step forward from oil or propane forced air systems. But how long will natural gas be a cheap option as carbon curbs are introduced?

To get off fossil fuels completely and as cheaply as possible, an air-to-air heat pump with an on-demand tankless electric water heater will provide heat and water as needed but at a higher cost than a natural gas system.

However, stepping up to a geothermal system should reduce annual costs even more than a natural gas system. The initial investment may be far greater but so is the future proofing and resilience. An all-electric system could allow for vehicle-to-grid electricity flow which facilitates your EV being your backup electrical supply should the grid go down for a period.

To put this in perspective, in early 2022, installing 3 Tesla PowerWalls (so named because they are wall-mounted batteries) with 13 kWh storage capacity each would cost approximately $35,000 in North America and give the home 39 kWh of battery backup. A base 2022 Nissan Leaf with V2G (vehicle-to-grid) capability has a battery size of 40 kWh and costs $44,000 before government incentives. The 62 kWh version of the Leaf sells for $47, 000. On any level a 55% increase in battery size for an additional 7% in cost would seem to be the obvious choice.

As an added bonus when comparing the Leaf vs the PowerWalls, the Leaf comes with wheels and virtually installs itself.

Below is a list of upgrades with costs and time for each for comparison purposes only. Your time and costs may vary (a lot) (Table 8.1).

Table 8.1 Cost and Impacts of home upgrades

Upgrade	Time	Cost (for scale purposes only - get local quotes)	Energy impact per $ 1–10	Carbon impact/$ 1–10	Comfort improvement impact 1–10	Resilience improvement 1–10
Window film	Hours	<$100	10	10	4	2
Door sealing and window caulking	1 or 2 days	<$200	10	10	4	2
Storm doors	1 day	$700/door	7	6	2	2
Storm Windows	Several days	$ Several thousand	7	5	3	2
Ceiling insulation upgrade	1–2 days	~ $4000	8	6	4	4
Installing granite countertops in the kitchen	3 days	~ $10,000	0	0	4	0
New Windows	1 week	$10,000 +	5	4	4	4
External roll-down shutters for large windows	Several days	$2500.00/ window	4	3	5	5
Tankless water heater	2–3 days	$4000 +	4	3	2	2
Natural gas boiler with hot water radiators	One week or more	~ $15,000	5	4	4	1
In-floor heating	weeks	$$$	4	3	7	1
Air-to-air heat pump and mini-split head units	3 or 4 days	$7000– $15,000	7	6	4	4
Geothermal heat pump	1 or 2 weeks	~ $35,000	8	8	2	6
Geothermal heat pump, solar hot water panels	2 or 3 weeks	~ $60,000	9	9	2	8

(continued)

Table 8.1 (continued)

Upgrade	Time	Cost (for scale purposes only - get local quotes)	Energy impact per $ 1–10	Carbon impact/$ 1–10	Comfort improvement impact 1–10	Resilience improvement 1–10
Geothermal heat pump, solar hot water, solar PV panels and 40 kWh of battery storage	weeks	~ $120,000	10	10	2	10

Clearly, the net effects of the above upgrades are dependent on what was installed to begin with. Here it is assumed that the starting system is a 70% efficient natural gas forced air furnace. There is a groaning smorgasbord of options in this field and those listed above are but a butter-plateful. In broad terms, upgrading a house to reduce energy demand by 80% and carbon emissions by 90% can cost in the range of $40/square foot in North America, 800–1000 pounds per sq. m in the UK and 1000 Euros per sq. m in Europe.

For a much more comprehensive look at the range of options, please see the referenced links.[4]

Windows play a large but passive role in the energy consumption profile of a house. Upgrading windows is low but very effective tech.[5]

---

[4] Retrofitting to Passive House standards

- https://www.youtube.com/watch?v=wN34zF7e4J8
- https://architecture2030.org/

Immense database on all aspects of environmental and energy issues from single home to district and national scale. http://www.2030palette.org/ In English, Chinese and Spanish.

https://passivehouseaccelerator.com/ Scores of passive house and building project case studies

[5] http://www.greenspec.co.uk/building-design/windows/

Window heat gain calculator https://susdesign.com/windowheatgain/

https://homeguides.sfgate.com/calculate-heat-loss-through-windows-26110.html

Calculate the cost of heat lost per square foot of the window by multiplying your fuel cost per unit by the number of degree days. Multiply the result by 38.82 for electricity, 1.57 for oil and 2.03 for natural gas and divide the result by 10,000. The end figure is the cost of heat lost per square foot of a double-paned window. For example: Cost of electricity: 98 cents, multiplied by 5000 DD equals 4900, multiplied by 38.82 equals 190,218, divided by 10,000 equals $19.02 per square foot of window annually. For triple-paned windows, multiply the result by 0.65. For a single-glass window, multiply the result by 2.27.

https://www.alulux.com/roller-shutters/solutions/

https://solariscanada.com/security-solutions/roll-shutters/

- Windows can both lose and harvest a great deal of energy.
- External blinds can control solar load and also reduce heat loss like storm doors.
- Internal insulating blinds greatly reduce heat loss.
- Caveats abound (Fig. 8.5)[6]

Fig. 8.5 Insulating cellular blinds. Two layers of cellular material plus 2 cm airgap produces around an R7 value

---

[6] Be aware of the LARGE CAVEAT on trapping heat between blinds and inner layer of glass. If the sun strikes a window with a highly insulating interior blind there may be a buildup of heat between the blind and the inner layer of glass causing it to become much hotter than the outside layer of glass. The inside sheet of glass expands and stresses the window seals and can even shatter. It is best to have windows with some sun exposure equipped with exterior roll-down shutters if they have highly insulation interior curtains or blinds.

Case Study: A Practical Example

A typical 2200 square foot (220 m²) house built in the 1990s in a moderately northern climate at the level of Toronto, Minneapolis, Stockholm, Sapporo or Minsk. It uses both natural gas and electric and is being converted to electric only running a heat pump system for space heating and domestic hot water.

Currently the house consumes an average of around 800 kWh of electricity per month or 10,000 kWh per year at an all-in cost of 20 cents per kilo Watt hour. Additionally it consumes an average of 300 cu m of natural gas monthly at a cost of 38 cents per cu m.

The current annual energy bill of this house is $2000 for electricity and $1400 for natural gas for a total of $3400.

A cubic metre of natural gas has the energy content equal to about 10 kWh. At a price of 38 cents, it is much cheaper than electricity on a cost per energy content basis coming in at 3.8 cents vs 20 cents, roughly a factor of 5 times less. But although natural gas is the cleanest and most efficient fossil fuel, it is still less efficient than electricity. And, electricity can be used to do work that fossil fuels simply cannot. A heat pump does not produce energy, it harvests and transfers it. Hence where fossil fuel efficiency tops out as it approaches 100%, electricity driven heat pumps are only just starting to clear their throats.

If the home owner simply switched from natural gas heating to resistance electric heaters like baseboards, the energy bill would be the same $2000 for the current electrical usage plus $7000 ($1400 × 5) to replace the energy produced by natural gas using the 5 times greater cost per energy content factor noted above. The total energy cost would be $9000 annually.

Enter the heat pump. At an average Coefficient of Performance (COP) of 4:1 for an air-to-air heat pump in that mid-northern climate regime, the cost of electricity used for heat would drop from $7000 to $1750 (7000/4) for a total annual bill of $3750. Payback on a $25,000 heat pump with three air handler units (min-splits) would be 5 years. This doesn't include any reduction in air conditioning costs, which would likely be substantial.

If the home owner opted for a geothermal system where the COP might be in the neighbourhood of 10:1, the annual energy bill would likely be around $2500 or less accounting for some air conditioning savings. The cost of adding the in-ground geothermal loops or boreholes might be $15,000–$25,000.

Clearly, if the combined geothermal/heat pump investment needed was $50,000 at the top end and annual energy costs (now all electricity), also came in at the top end of $2500, down from the current energy bill of $3400, it

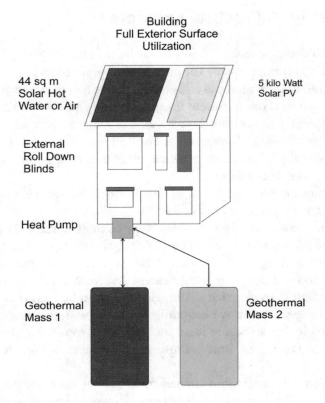

Fig. 8.6 An energy optimized house

would be a daunting prospect from a strictly financial payback point of view (Fig. 8.6).

From a carbon emissions point of view, however, it would be a massive improvement. A cubic metre of natural gas emits 1.86 kg of carbon dioxide when burned. The carbon budget of the house based on natural gas alone (3600 cu m per year) would then drop from 6.7 tonnes to as low as zero if the grid is entirely renewables powered.

The geothermal heat pump system is expensive at the beginning but it then becomes the core of a system which can grow in a number of different directions. A Tesla PowerWall and solar PV array could be integrated seamlessly. These would both greatly reduce electricity costs. An EV with V2G capability could also power the house when the grid goes down or rates are high.

Solar hot water could be added to the system easily and by storing heat and/or cold in the ground could increase the COP above the 20:1 level. The Drake Landing development in Okotoks, Alberta with 52 housing units is a full-on borehole geothermal storage system and it is seeing Coefficient of Performance levels close to 30:1.

At that point, energy costs fall solidly below those of any fossil fuel alternative. The front end investment is larger and maintenance and insurance[7] costs may increase slightly but the environmental costs fall dramatically and well into the realm of planet-saving levels. And then there are the benefits which accrue off the cash journal.

Resiliency increases dramatically and energy cost stability is greatly enhanced. Many people will see their real disposable incomes eaten up by rising energy costs. Owners of houses upgraded to the levels above won't be among them.

A 10 kWh photovoltaic solar panel installation on the roof of this house would likely produce (at this latitude) about 12,000 kWh annually. With a few additions like roll-down shutters and insulating blinds, this house could produce a surplus of electricity of 4000 kWh annually, which is enough to power an efficient EV 25,000 km. The more energy you save in the house, the more gasoline you can replace on the road.

A note must be made here on grid storage. Grid storage can occur when a house generates more electricity than it needs and sends it into the local electricity grid. Of course, the local utility has to agree to this arrangement but most will most of the time.[8] But once they do, they will give you a credit for the power you give them and this can be redeemed at any time.

However, they are providing a service so a storage and handling levy will be applied. This can typically be around 50%. So if you transfer 10 kWh of electricity into the grid and your typical cost from the grid is 20 cents per kWh, you will be able to use that 10 kWh at any time but the utility will charge you 10 cents per kWh for the storage service.

By 2050 most homes will be running strictly on electricity and be equipped with at least some of the systems discussed above. The key point is that once the commitment to a geothermal system is made, many options for both energy harvesting and storage become available. A very efficient, diverse and integrated system can be built over time in progressive steps.

Hanging on to a fossil fuel based system for as long as possible limits good options and may force owners to do the complete upgrade over a very short period of time.

---

[7] In terms of insurance, there is a trade-off between the increased cost of the new system vs the elimination of natural gas from the property. Discuss this with your insurance agent.

[8] The utility has to be sure that its network can handle the power you (and potentially many other home owners) may be transmitting over their high tension lines. Talk to them before making assumptions.

As with Covid-19, it is better to decide on a path early and then get on it as hard as you are financially able. Then you can pick your options from a much broader palette rather than being dragged through a succession of half measures.

Comfortable, Healthy and Efficient

Which is the most comfortable type of heating system? Is it forced air where a different temperature of air is blown into the living space? Is it electric heaters which create columns of rising hot air with return flows of cooler air? Both of these tend to create cold floors.

Most people very much enjoy woodstove and gas fireplace heat that radiates into the room, creating an area where all of the surfaces are warm to the touch. This kind of heat can create cold spots in the house despite the very pleasant feel of the "warm zone". Good insulation can go a long way to making any heating system comfortable by eliminating drafts and cold spots in the house.

But the most enjoyable system is generally thought to be in-floor heating where the heat rises evenly through the room and the temperatures of most surfaces are very close. And walking on a slightly warm surface reduces stress. This adds to the effect of in-floor heating feeling of warmth and comfort.

Besides its feel, in-floor heating has the potential to be the most efficient way to heat a house. However, its installation costs are the highest and retrofitting is particularly expensive compared to installing hot water radiators or mini-split air handlers. These produce comfortable heat but just not up to the level of in-floor.

Building New

The advantages of building new are tremendous, especially if it is possible to select a building site that will allow the home to be oriented for optimum energy harvesting and minimum heat loss. If the house used in the Practical Example above were to be built today, it would also enjoy the benefit of 30 years of substantial technological advances.

The advances have occurred not only in stand-alone components but in their ability to complement each other and integrate energy saving and energy harvesting in synergistic designs. These cut consumption, increase comfort and, in the long run, reduce costs.

Roof design is critical (Fig. 8.7).

Here on the left we see the style of house built in the early twenty-first century with numerous roof surfaces serving no productive use and rendering the house extremely expensive to upgrade for solar panel installation. Compare this to the house on the right, oriented nearly dead south with 100% of its roof surface available for energy harvesting systems. Currently it houses a solar hot air system but solar hot water or photovoltaic panels (PV) could easily be added. When designing a new house, energy harvesting and orientation must trump style.

Thermal mass and insulation can be built into the walls, the foundation and the pad in ways that could never be retrofitted to an existing building. Window wells can be designed to accommodate external shutters and interior insulating window treatments.

Heating and cooling systems, particularly in-floor types, can be built into new homes vastly more easily than they can be added to existing ones. Drilling or excavating the property for the installation of geothermal borehole or ground-loop piping is easier to do when there is no existing house to work around.

Room specific heating control and thermal management can contribute to both lower energy consumption and greater comfort. This is a large subject and best discussed with your architect but is described in more detail here[9] at

Live in the style of bygone days in an unfixable house or live comfortably in a house built for the future

Fig. 8.7   House designs we are building vs. house designs we need to build

---

[9] https://passipedia.org/basics/building_physics_-_basics/thermal_comfort/local_thermal_comfort

the Passipedia, Passive House Resource. The best way to end with an energy positive house is to start with a passive house.[10]

Once you and your architect have given the house "good energy bones" by way of siting, orientation, roof and window and material choices, it will be up to your builder to assure that the insulation and seal fulfill the potential of the design.

In all, new houses south of the 50th parallel, located well inland should be able to be made energy positive while houses near the coast should be able to achieve that status at the 60° level. Numerous caveats on local wind and humidity factors apply but, with a clean slate, energy positive houses are feasible in the areas where the great majority of people on this planet live.

Geothermal Overview: Technology that Deserves a Spotlight

Due to the losses inherent in transporting low grade energy like heat, geothermal systems are local systems. They are either on an owner's property or very nearby as part of a district heating system. When solar hot water panels are added, these will go either on the home or very close-by.

On the positive side, low grade energy is much easier to collect than the high grade energy, electricity. PV panels are currently able to convert around 20% of the suns energy into electricity. Hot water panels now exceed 50% efficiency levels and they are cheap and easy to install.

The critical job that needs to be done in northern countries is the storage of energy over the 4 or 5 winter months when solar energy availability falls far below the daily heating needs of most buildings.

Geothermal heat storage is extremely cheap with its cost per kWh of energy being less than 1% of that of electrical battery storage. And, of course, heat has been taken out of the ground around the boreholes or tubes, the earth will endeavour to bring the temperature back up to ambient ground temperatures.

- 10 https://www.youtube.com/watch?v=q7WWHhcMCdUPassive house institute
- https://passivehouse.com/
- https://passipedia.org/
- Passive house instruction manual - https://passivehouse.com/05_service/03_literature/030300_user-manual/030300_user-manual.htm
- How is your German?? Nicht so gut? Then you can still get through this in reasonable shape by translating the pdf text using Google translate (www. https://translate.google.com)

In the winter, the ambient air temperature will almost always be lower than the ground temperature and air-to-air heat pumps will therefore have to work harder to extract heat. Fig. 8.8 shows the COP difference a heat pump will experience between air and ground sources. On a winter's night of –5 °F in Chicago, the ground is 51 °F or 30 °C degrees warmer than the air temperature (Fig. 8.8).

The impact this has on energy efficiency can be seen by looking at the COP of 12 for a 15 °C degree difference vs a COP of 4 at a 45 °C degree difference. In other words, for every 1 kWh of energy put into a heat pump at a COP of 4, it will output 4 kWh of energy into the house whereas a COP of 12 will output 12 kWh or 3 times more heat.

It is easier to squeeze heat out of warm earth than cold air.

On the map below the ground temperature difference between Atlanta, Georgia (at 67 °F) and Ottawa, Canada (at 37 °F) is 30 °F or 17 °C. A 17 °C difference equates to about a doubling of the COP, leaving Ottawa home owners to pay twice as much as their Atlanta counterparts for the heat they produce. Additionally, Ottawa residents need a great deal more heat over the course of a winter than do Georgians.

The energy required to heat a home in the north can easily amount to 80 kWh per day of winter. The chart below shows what an impossibility that represents for electric batteries. Contrasted to our old fossil fuel system great strength of built-in storage with a 1000 litre oil tank having 10,000 kWh or

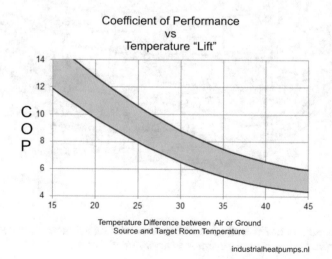

Coefficient of Performance
vs
Temperature "Lift"

Temperature Difference between Air or Ground
Source and Target Room Temperature

industrialheatpumps.nl

Fig. 8.8   The efficiency of a heat pump can be improved immensely by giving it higher temperature input, namely geothermal

125 days of heat supplies in it, a $10,000 Tesla PowerWall with 13 kWh capacity could only provide 4 h of heat through electric baseboards.

But used to drive a geothermal heat pump that 13 kWh could provide about 156 kWh of heat at a COP of 12 which would meet 2 days of heat requirement (Fig. 8.9).

Despite the fact that the earth will always be trying to return ground to its natural ambient temperature, as heat pumps with draw heat, ground temperatures fall. By the end of a hard winter in a northern region, the ground temperature may well be close to zero. The earth has all of the summer and fall to restore temperature levels but this process means that the COP of a geothermal heat pump will be higher in the early winter than in the early spring.

Art Hunter who runs a living lab (description in a later chapter) on his property has the following comment: "I can speak from experience with a GSHP (Ground Source Heat Pump) as the sole method of heating/cooling and heating domestic hot water, the energy savings are substantial. I measure a COP as high as 15 in the fall and down to a low of 4.2 at the end of February."

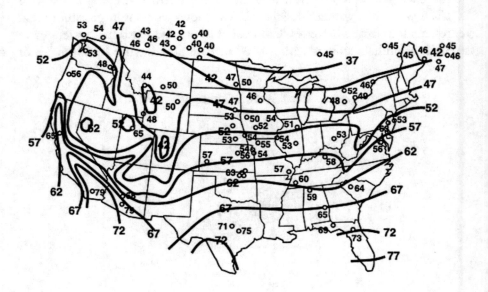

Mean annual earth temperature observations at individual stations, superimposed on well-water temperature contours.

Fig. 8.9 The starting temperature for your geothermal heat pump system is the ground temperature in your region

Geothermal can be used for air conditioning as well as heating. Australian GSHP systems can take advantage of ground temperatures of between 15 °C and 18 °C in Sydney and Canberra, 23 °C in Brisbane but the 31 °C in Darwin might make air-to-air a better investment.[11]

Geothermal Heat Pumps

By compressing or expanding the gas in the condenser or evaporator coil in a heat pump, it is possible to make that section of the tube hotter or cooler. Air (or possibly liquid) is then passed over the hot or cold portion, transferring the heat to the outside air or reducing its temperature. Heat pumps do not create energy; they transfer energy from one area and one material to another.

Hence it is possible to transfer heat from the ground or lake or outside air into the house or to expel excess heat from the house. Refrigerators expel heat from their interiors into the kitchens air mass.[12]

That is the means by which a heat pump transfers heat but in rare situations it is possible to deep drill and tap into sufficiently hot ground to produce electricity. Heat flows above the 180 °C temperatures will spin a steam turbine with sufficient force to generate electricity. The Hellisheidi geothermal energy plant 30 km east of Reykjavik produces 300 MW (300,000 kW) of electricity and the equivalent of 400 MW of heat energy. This heat is piped into the city centre and used to heat buildings and sidewalks to melt ice and snow during the very long Icelandic winter.

These projects are not as simple as they might seem with hot fluid being extracted through 30 wells at a depth of 2000–3000 m spread over an area of 13 hectares. Care must be taken to assure that energy flows are not over-exploited leading to permanent loss of heat transfer capacity due to altered magma flows (Fig. 8.10).

Iceland and other regions blessed with the double edged sword of active volcanos can make these systems work with drill depths of 3 km but in other areas where the technology is potentially practical, such as parts of Europe, well depths can exceed 6 km.[13]

---

[11] https://www.jstor.org/stable/sanctuary.14.72?seq=1

[12] https://www.youtube.com/watch?v=-r6L9HPJfnE
   https://www.youtube.com/watch?v=1jCHYUuEDZ8

[13] www.euronews.com/living/2020/11/28/fire-and-ice-geothermal-energy-s-world-saving-potential
   www.youtube.com/watch?v=nnqdUQo7cWQ
   www.builditsolar.com/Projects/Cooling/EarthTemperatures.htm

How a Heat Pump Works

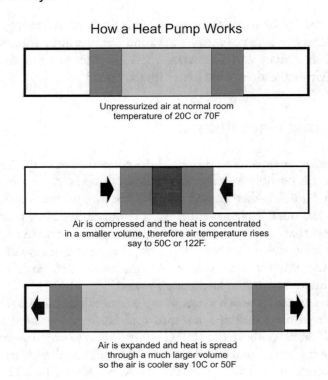

Unpressurized air at normal room
temperature of 20C or 70F

Air is compressed and the heat is concentrated
in a smaller volume, therefore air temperature rises
say to 50C or 122F.

Air is expanded and heat is spread
through a much larger volume
so the air is cooler say 10C or 50F

Fig. 8.10   A heat pump does not generate heat, it transfers it

Unlike the extremely accessible ground source heat pump, costs can be high and difficulties great with geothermal power production and it will likely remain an option few areas can practically exploit.

Geothermal Benefits Summary

Geothermal heat pumps require a sizable initial investment but they also deliver the most heat or cooling capacity for the lowest energy input. They have the lowest inherent carbon footprint and form the basis for an open architecture of energy flows on the property and for connection to other properties by way of heat micro-grids or district heating.[14]

---

[14] www.youtube.com/watch?v=PI45yUhUWgk

Pros and Cons of Mini-Splits/Ductless

Pros

- No ductwork, minimal space requirement
- control temperature in each zone independently
- heat pumps are air conditioners as well
- more efficient than ducted systems
- major mechanicals are placed outside the house envelope

Cons

- higher initial cost
- reduced low temperature performance in an air-to-air system may need resistance heater augmentation below -15 °C. (no problem with ground source)

Good overview: www.hgtv.com/design/remodel/mechanical-systems/is-ductless-heating-and-cooling-right-for-you

To Drill or Not to Drill, Borehole or Trench?

If your intention is to eventually move to heat storage in your geothermal system, then the further the system is below grade, the better. Boreholes require less area but can, depending on their depth, access greater thermal mass. A deep heat mass will retain a high temperature better than a one only several feet below the surface. Is there a significant flow of subterranean water at the depth you are going to be drilling? If so, then perhaps heat storage would not be any better than with a trenched system.

Excavating equipment is readily available whereas drilling boreholes is more specialized. Discuss this with your heat pump system installer.

Cold geothermal storage may not work as well as heat storage but it certainly can work well enough to reduce energy consumption in regions which are subject to high temperatures. In the United States, heating consumes 15% of electricity production while air conditioning consumes 16% (Fig. 8.11).

An energy positive house does not produce surplus energy all the time but rather stores energy during periods of surplus and then withdraws energy from storage in periods of deficit. Relatively small amounts (several days of winter requirements) of electrical energy can be stored in chemical batteries such as Tesla PowerWalls or EVs. Surplus electricity can also be fed into the grid for withdrawal at a future point in time.

Heat energy can be stored underground and, if the grid service charges are too high, electrical energy can be converted to heat and stored underground for withdrawal months later. A geothermal system of say 10 boreholes spread

Your 100% Electric House as Energy Positive

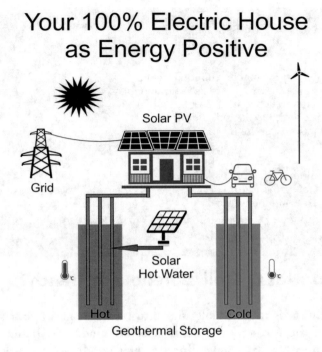

Solar PV

Grid

Solar
Hot Water

Hot

Cold

Geothermal Storage

Fig. 8.11 House energy systems in a fully connected house

over 100 m² with a depth of 30 m would access a thermal mass in excess of 6000 tonnes of rock and soil. The energy storage capacity, in the form of heat, of such a system is far greater than any conceivable electric batteries that could be built into a residential energy system. More on this in Chap. 10 "Your Community".

Fast forward to 2050 and the archetypical home in the western world may or may not be a single family dwelling but it is most certainly powered by electricity. It consumes a great deal less energy than those of one or two generations earlier and it now produces energy, possibly even more than it consumes. If the family has a vehicle at its exclusive disposal, it is electric and the household may also have ebikes as well.

This residence will be connected to electrical and thermal energy systems which will include both generation and storage capacity in myriad forms. The recycling "bins" in the home will allow 100% separation of material flows to facilitate a very high level of recycling at the district centres.

Compared to our life today, we will be much more aware of the material and energy flows around us. Most of the complexity of these flows will be handled automatically but life will still more resemble that of our hunter-gatherer forefathers than the 1950s super consumer ideal of complete resource environmental detachment.

Solar PV and Hot Water Practical Notes

These systems are ground or roof mounted. Ground mounting takes up perhaps valuable space while roof space can be thought of as free. The racking for roof mounting is less elaborate than for ground mounts but access may be more difficult. For roof mounts, it is better these systems go on steel roof surfaces than shingles since steel has a longer lifespan. Having to remove the entire installation to replace asphalt shingles has been a very expensive proposition for those who have had to do it.

In order to install a PV system on your property and tie it into a building approval must be granted by the local electrical utility. They have to assure that the safety of their grid is maintained and also that their grid will be able to handle the power your system has the capacity to produce at the time it is likely to produce it.

Solar panels vs solar shingles and tiles[15] a few points to consider:[16]

- Solar panels are a great deal cheaper and they their efficiency is higher. In early generations of solar shingles, efficiency was about 1/3 that of panels. Hence shingle producers were loath to quote efficiency.
- But for solar PV, efficiency is as key a specification as horsepower was to American Muscle cars or range is to EVs. If the shingle supplier you are speaking with can't quote efficiency ratings off the top of their head, just walk away.
- Shingles/tiles have become more efficient and don't require a re-build of the roof into a flat plane but their installation takes more time with an experienced crew.

Solar Benefits Summary

- Invest in your own energy and control your own supply and future costs
- Minimize peak power rates
- Drive your EV "for free"

---

[15] https://tbsspecialistproducts.co.uk/pv/
[16] https://www.youtube.com/watch?v=TxscCFyYPto

- Tie into community systems
- Tie into district geothermal systems
- Good background video on Fully Charged[17]
- Keep ahead of trends - the living lab outlook.[18] *Making decisions in an environment when staying the course is no longer an option is a lesson we should have learned from the pandemic.*

Solar Panel Pricing in Different Climate Regimes (Table 8.2)

Opt for the most efficient solar panels when space limitations put a priority on efficiency. Otherwise high panel efficiency might be an unnecessary expense. Efficiency is gradually improving but not quickly enough where it would pay to "wait for the next breakthrough" (Fig. 8.12).[19]

Your House and the Grid: A to B No Longer

The grid of the future will be more like a living organism with two way flows between many different types of energy generators, storage facilities as well as between energy suppliers and consumers (Fig. 8.13).

Note the large and diverse energy storage potential which will be able to eliminate curtailment (wasting or turning off energy generation that can't be used) from most renewable energy sources.

Table 8.2 Solar panel system price vary by region

Region	Cost per watt ($/W)	Solar panel cost range (6 kW system)	Solar panel cost range (10 kW system)	Capacity factor (%)	Cost per kW capacity
Arizona, USA	$2.34	$12,900–$15,180	$21,500–$25,300	30	$7.80
Connecticut, USA	$2.81	$15,240–$18,480	$25,400–$30,800	18	$15.60
Yukon, Canada	$4.50	$27,000	$45,000	7	$64.00

---

[17] https://www.youtube.com/watch?v=uI7wxNtrorQ
[18] https://microgridknowledge.com/siemens-living-lab-microgrid-princeton/
[19] https://www.youtube.com/watch?v=Y2qHeRy6CY8

Most Efficient Solar Panels 2021				
Brand	**Model Number**	**Efficiency %**	**Price per Watt $/W**	**Type**
Spectrolab	XJT Prime	30.7	5	Multi-junction
LG	LG395N2T	18.7-24.3	0.74-0.57	Bi-Facial Mono
SunPower	X22-370	22.7	0.95	Monocrystalline
REC Solar	Alpha 380	21.7	0.8	Hetrojunction
Canadian Solar	CS3W410P	18.6	0.62	Polycrystalline
Solar Frontier	SFK185-S	15	0.88	CIGS

* Prices may vary based on your region and amount of panels purchased

Fig. 8.12   Price per Watt generated and efficiency varies significantly

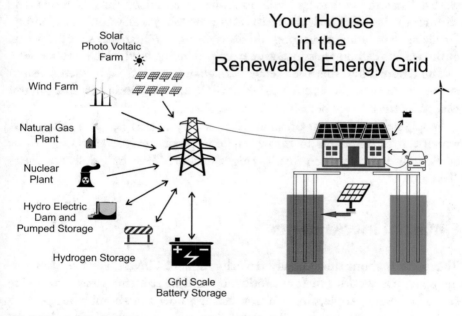

Fig. 8.13   A fully fleshed out renewable energy grid

Cottage

You will never be successful in getting away from it all if you take it all with you. Make the few toys you can't live without electric and get back to nature on nature's terms as often as possible. An EV with V2G capability would allow a cottage to be operated with no grid connection.

Travel

Travel is one of the pleasures of modern life but it comes with a fairly hefty carbon price tag. Reducing travel and travelling to destinations via high-speed train or your new EV rather than in a jet will dramatically reduce your carbon footprint. So will taking one longer vacation in place of a number of short ones.

Food

Dairy and meat products require many times more energy and resource inputs than do plant based foods. Eat lower down on the food chain and both you and the planet will be healthier.

The UN report[20] estimates 17% of the food produced globally each year is wasted. That amounts to 931 million tonnes (1.03 billion tons) of food along with all of the fossil fuels, materials and processing that went into planting, fertilizing, harvesting, packaging and distributing it. Apparently sizable fraction of the world's carbon emissions were from producing food which was wasted.

The report found that the average Canadian wastes 79 kg of food a year at home, more than the average American (59 kg) and similar to the amount wasted by the average person in the U.K. (77 kg).

People need to be clear on labels like "Sell By", "Best By" and "Enjoy By" which can cause people to throw out food prematurely even though some labels only indicate when quality might decline. "Best by", does not mean "bad after".

Financial Investments

This book is about substantially replacing financial investments with investments in real wealth (energy) production and in reducing long term living costs. To many people, financial markets are now time-bombs but picking peaks or winners/losers is well beyond the scope of this book as well as the psychic predictive capacity of the author. But one telltale sign of an on-coming dramatic change might be the cessation of inflationary money printing.

Asset inflation is rampant in today's economy of money-printing and "wealth effect" Ponzi schemes. Some will profit from these zero sum market

---

[20] https://www.unep.org/resources/report/unep-food-waste-index-report-2021

machinations but most will lose. Maybe this is not the train to be on when biophysical impacts start to be broadly felt.

A stock market crash won't stop your solar panels from brewing your morning coffee. But you will need to have installed the solar panels before the crash.

Climate Change: Coming Soon to a Town Near You

Unpredictable Points to Ponder

Reports: Flooding risks could devalue Florida real estate
   Two new reports say flooding due to climate change-related sea level rising and the erosion of natural barriers pose substantial economic risks to Florida, particularly to the value of South Florida real estate

- Heading for the exits – when the tide turns, how quickly will markets drop for flood (fire prone, climate prone) prone real estate, gas power vehicles and???
- Don't be caught in a stampede
- Is sustainability now a thing?
- "We surveyed 20-odd developments in Perth and found that location was the number one thing buyers were concerned with."
- "The second thing was the price and the third thing was sustainability, and that came even before how it looked. They really wanted to know a lot about the energy efficiency."
- https://abcnews.go.com/US/wireStory/reports-flooding-risks-devalue-florida-real-estate-68435458?cid=clicksource_4380645_9_heads_posts_headlines_hed
- https://www.abc.net.au/news/2020-01-23/home-design-energy-efficiency/11886208

Perhaps surprises will be the only predictable aspect of climate change. Extreme weather events are more likely to occur now than during the period of the past 200 years of "tranquil climate" during which we began to collect reliable data. In the middle and late 1900s we depended on those weather records to inform the design and placement of our social infrastructure. But we will now have to admit that we are substantially blind to future trends and will have to place much greater emphasis on resilience engineering than pushing the "limit" which has now become a nebulous and moving target.

There are many facets to the same topic and the three graphics below illustrate how different an issue can look depending on the parameter being examined. "It never rains but it pours" is an old expression and one unlikely to lose

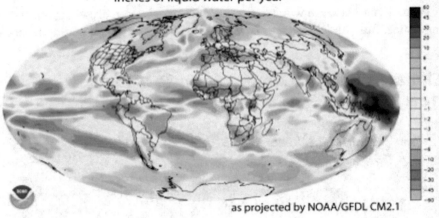

CHANGE IN PRECIPITATION BY END OF 21st CENTURY
inches of liquid water per year

as projected by NOAA/GFDL CM2.1

Fig. 8.14 Changes in annual precipitation by 2100. NOAA

validity going forward. Rain intensity is predicted to increase and, with surface water evaporating more quickly once the runoff is over, so may aridity.

There are two parts to the precipitation story: danger from floods and drought. It is possible to have more rain yet more severe droughts as rain is more concentrated in a shorter period and dry periods are extended.

Pick your geographical location carefully and choose the position of your home relative to waterways and potential flood areas even more carefully.

There is rainfall, there is rain intensity and then there is evaporation. Low annual rainfall no longer means low runoff or immunity from floods or shifting ground (Figs. 8.14, 8.15, and 8.16).

Pick Your Spots: Literally

Whether you live in a house, condo or apartment or in the city or in the country, the region in which you live has a great deal to do with the size of your energy and environmental footprints. Moderate climates typically have the lowest footprints, the healthiest lifestyles and the greatest access to locally grown food. The book "America's Most Sustainable Cities and Regions; Surviving the 21st Century Megatrends"[21] by authors John Day and Charles Hall deals with a good many of the questions involved in making a sound strategic choice.

---

[21] https://www.springer.com/gp/book/9781493932429

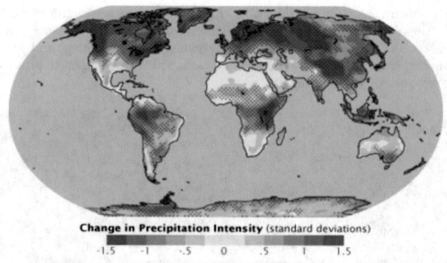

Change in Precipitation Intensity (standard deviations)

-1.5   -1   -.5   0   .5   1   1.5

Precipitation intensity: a larger proportion of rain will fall in a shorter amount of time than it has historically. Blue represents areas where climate models predict an increase in intensity by the end of the 21st century, brown represents a predicted decrease. NASA Earth Observatory

Fig. 8.15   Change in precipitation intensity by 2100. NOAA

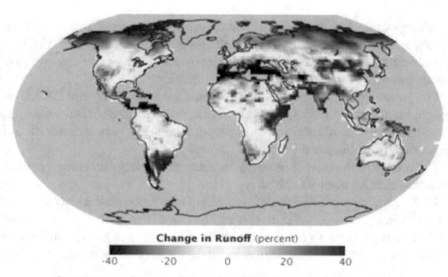

Change in Runoff (percent)

-40   -20   0   20   40

Changes in water runoff into rivers and streams are another expected consequence of climate change by the late 21st Century. This map shows predicted increases in runoff in blue, and decreases in brown and red. (Map by Robert Simmon, using data from Chris Milly, NOAA Geophysical Fluid Dynamics Laboratory. NASA Earth Observatory

Fig. 8.16   Change in volume of runoff or water flow in rivers by 2100. NOAA

Similarly, the regions in the world which have the best temperature for the development and maintenance of civilization are measured by the same yardstick. Can you re-locate to one of these regions?[22] Will technology and your own expertise allow you to prosper in an area which would seem to present disadvantages to most people?

Cheap and Clean Decarbonization List

- Make your next vehicle an EV,
- walk and bicycle instead of driving as much as you can,
  - maybe an electric assist bike would make it possible to avoid using a car for some medium distance trips
- install a ground source heat pump for your house and business
- electrify your lawn mower, snow blower, garden tools and chain saw
- eat local food as much as possible
- wear more clothes and turn down your thermostat
- buy quality products which will last a long time and which can be repaired, upgraded and recycled

So It Is Up to You

Speaking out against institutional corruption is a long term and likely frustrating activity but directing your own life towards the goal of sustainability is not. There are a great many things which you and your family can do today which will deliver very positive results very quickly.

Put yourself in the position of being able to measure energy. "With an energy management philosophy you find it easier to understand resilience, sustainability, adaptability and mitigation. These are very difficult if not impossible to address in the world of money."[23]

Use money in the market but base decisions and strategy on energy. Money costs will change, energy math won't.

In the early days of fossil fuels it was cheaper to just throw energy at whatever problem presented itself; to just hit the throttle as it were. Now that we are well past peak cheap energy, it is cheaper to reduce demand than it is to produce more energy. Conservation has always been the key to sustainability.

---

[22] https://www.youtube.com/watch?v=oG19fCFSamQ
[23] Art Hunter.

Lowering demand and increasing the harvesting of energy is the equation that will produce the lowest cost, healthiest and most secure way of living while remaining both mobile and prosperous.

> *Do what you can, with what you have, where you are.*
> Theodore Roosevelt

9

Cars, Tools and Mobility

Introduction

There is now a massive number of Evs coming on the market. Given the issues they bring with them as well as the technology they contain, it would be possible to write a book a month on the subject of the transition of ground transportation systems from fossil fuel to electron motivation. This chapter is about the progression of development and the categories of products to allow the reader to better understand their own needs and determine the products which best meet them.

Many of the vehicles listed here are typical of a category. The most important piece of information you can have is a clear idea of what your needs are so you can quickly identify the category most suitable for you.

So here, in this single chapter, are the basics. Battery electric vehicles (BEVs) – I'll use the term "EV" in this book, – are massing on the hills above the hapless wagon train of internal combustion engine (ICE) vehicles and are preparing to sweep them into history. The cavalry won't be coming over the hill in this case because, actually, EVs are the cavalry.

EVs Versus ICE

If efficiency, reduced carbon and dwindling cheap oil are the future, then EVs are the future (Fig. 9.1).

A great deal of energy goes into producing, refining and distributing oil and gasoline. The natural gas production is 95% efficient. When the 950 kWh

J. E. Meyer, *The Post-Pandemic World*, https://doi.org/10.1007/978-3-030-91782-1_9

Oil Sands - Internal Combustion Automobile Lifecycle

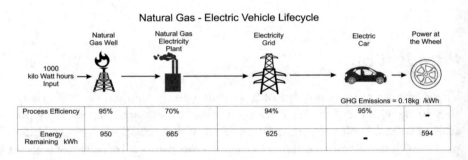

	Natural Gas Well	Oil Sands	Bitumen Upgrader	Oil Refinery	Oil and Gas Distribution	Internal Combustion Engine Car	Power at the Wheel
1000 kilo Watt hours Input →							GHG Emissions = 1.06 kg / kWh
Process Efficiency	95%	72%	88%	88%	93%	35%	
Energy Remaining kWh	950	3325	2927	2575	2394		838

Fig. 9.1 Fossil fuels involve many more steps, more ghg emissions and more energy than an electrical system

Natural Gas - Electric Vehicle Lifecycle

	Natural Gas Well	Natural Gas Electricity Plant	Electricity Grid	Electric Car	Power at the Wheel
1000 kilo Watt hours Input →					GHG Emissions = 0.18kg /kWh
Process Efficiency	95%	70%	94%	95%	-
Energy Remaining kWh	950	665	625	-	594

Fig. 9.2 Natural gas sourced electricity in EVs is far better than internal combustion vehicles but solar panels on your roof are the best solution

of natural gas is used to produce oil from the oil sands the total energy available grows to 3325 kWh but from there each process reduces energy content. In the end process, the vehicle motor efficiency is only 35% meaning that only about 1/3 of the energy contained in the gasoline is translated into motion. The rest goes up in heat.

Below is the lifecycle of an EV using natural gas as the base energy to drive an electrical plant. The energy consumed and the emissions created are vastly lower than for a natural gas - oil sands - internal combustion engine chain.

Both energy consumption and ghg emissions would be much lower for a process driven by wind or solar, particularly solar panels on your roof (Fig. 9.2).

Why are electric motors so much more efficient than gasoline or diesel motors? As detailed in Chap. 5, the air pumping, friction and combustion heat losses exit the engine via the exhaust and radiator, causing efficiency to be much lower in an internal combustion engine. Fossil fuels have great strengths but motors which run on them aren't very efficient, possibly 35% vs 95% for an electric motor.[1]

---

[1] https://medium.com/swlh/internal-combustion-engine-and-the-four-stroke-engine-12381fc54a98

EVs: Where Needs and Wants Converge

Few technological advances have received the level and duration of hype that Electric Vehicles have over the past decade. EVs are The New Must-Have Product. How did that come about? As events unfolded, EV producers, led by Nissan and the much flashier Tesla, triggered the perfect storm of consumer enthusiast interest in the greater context of an environmental crisis.

There is a tsunami of information on EVs on the internet because they have been embraced by tech-heads, gear-heads, tree huggers, sports car drivers, health advocates and all manner of civic planners. They appeal to the technically minded and they solve a huge number of problems for cities and delivery and taxi businesses.

They offer a level of refinement previously unavailable to many buyers and they offer responsiveness, never previously available: period.

Tech heads love them for their new technology and the astounding amount of information they generate. Plus, in many cases, the operational controls were more similar to those of a computer than a traditional car. EVs have screens, information and configurability options which exceed those of a cell phone.

The second group, of course, are the automotive enthusiasts: people who take their driving seriously and are finely attuned to the operation and driving characteristics of their ride. EVs come to the table with a huge amount of baked-in driver appeal. They offered faster throttle response than any gas car is capable of and, having only one gear (mostly), they deliver a seamless surge of power from a dead stop. Internal combustion engines (ICEs) need to build up their revolutions per minute, to deliver both power and torque while electric motors deliver maximum torque at zero RPM.

Consequently, EVs are fast off the line and even pedestrian versions are faster than most of the V-8 heavies which thrilled drivers and passengers alike in the bygone muscle car era. And they deliver this performance with eerie levels of silence and smoothness sufficient to put luxury car makers on notice.

The third group were those committed environmentalists who wanted to reduce their carbon footprint and EVs delivered the biggest carbon reduction bang for the buck. Carbon pinchers were joined by penny pinchers who simply wanted to reduce their cost of driving. EVs are now cheaper in the long run to operate than gasoline and diesel cars and they require less maintenance and trips to the garage for service. And, if you have the ability to charge at home, you may never have the need to go to a gas (or charging) station ever again.

All of a sudden these new hot-rods/glitter-techno/earth-savers became the hottest consumer item on the planet in decades. Given the rapid advancement of technology and the high numbers of new models being announced monthly, I'll cover the basics plus, of course, some of the juicier tidbits.

Do EVs Have Their Problems? Yes

Some early EVs were close to the golf-carts-with-doors that detractors accused them of being. It is possible that early EVs didn't come with navigation systems because their low range made the possibility of going far enough from home to get lost well-nigh impossible. Current EVs however present absolutely abundant opportunities to get lost as well as arrested for speeding and stunt driving.

Range is still a problem for a small number of drivers and charging networks are not fully fleshed out in many regions. Also, EVs are simply a little bit different from petrol cars and "different" even if it is better, requires some adjustment.

Do EVs Have Their Detractors, Oh Yes! Or at Least They Used To

If you'd asked consumers in the late 1800s what they wanted as ideal transport they would likely have said a faster horse. Gasoline vehicles certainly had their problems and detractors in the early days. They were unreliable, gasoline was hard to come by, they were expensive, they smelled and they would occasionally burst into flames and explode. You could break an arm or even kill yourself trying to crank-start one and the road system was not car friendly. Initially they were mostly city vehicles partly because there was no gasoline available in the country.

Gas vehicles rapidly overcame their shortcomings and certainly the knocks against EVs have greatly diminished. Car aficionados who may have snickered a decade ago now probably own one.

There is a great deal of analysis of what isn't perfect about EVs but a good portion of that comes down to what is different about EVs.[2]

EVs and tire wear. EVs tend to be heavier than their ICE counterparts and so place more stress on the tires. Also, with their superior response and driving

---

[2] Electrifying, https://www.youtube.com/watch?v=FdxYXAXGcKw

characteristics, the demographic which has so far embraced EVs has tended perhaps to drive in a more spirited manner which also takes a toll on the life expectancy of tires. You mileage may differ but if you repeatedly indulge in that instantaneous throttle response, both your range and your tires will suffer.

A Short History of Electric Vehicles

The automotive world changed with the arrival of the Nissan Leaf. It was a smooth, fully developed, comfortable and spritely EV with close to 120 km or 80 miles of range. At that point electric cars became real. It took a while for this to set in to a bemused automotive establishment but by the time Tesla put its Model S on the market, greenness and practicality was shouldered off the stage by speed and flash. EVs had become an aspirational purchase. But let's begin at the beginning.

Ferdinand Porsche's first car was electric. After spending 108 years in storage, its engine still ran when powered up. The batteries were another matter though (Fig. 9.3).[3]

Note the early streamlining of the cowl and the hub motors in the front wheels.

In 1898, Ferdinand Porsche designed the Egger-Lohner C.2 Phaeton. The vehicle was powered by an electric motor, and with 3–5 horsepower it reached a top speed of 25 km/h. Two years later with 2 × 2.5 horsepower motors it

Fig. 9.3   An early electric Porsche

---

[3] https://newsroom.porsche.com/en/products/taycan/history-18563.html

reached a top speed of 37 km/h. Porsche's boss Lohner's reason for making a vehicle with an electric motor was that in some towns "the air was ruthlessly spoiled by the large number of petrol engines in use".

Petroleum reserve discoveries grew rapidly in this era and the internal combustion engine was the only practical means of getting an airplane off the ground. Hence, gasoline and diesel engines established almost complete dominance in the transportation field and battery development languished for decades.

The first Harley-Davidson had pedals to go with its gas engine. And now the company is offering both an electric bicycle and an electric motorcycle, the LiveWire (Fig. 9.4).

Please take a moment to contemplate the arrival of electric Harley-Davidsons. Let us all fall to our knees and give thanks for the coming end of the belching noisemakers which currently bear the company's logo.

Fig. 9.4   The first Harley Davidson bike had pedals – the serial 1

Three decades ago, the dawning of concern about global warming and the lifespan of oil reserves, interest in electric vehicles began a slow revival.

GM's EV1 in 1990 was the first credible EV from a major manufacturer with 137 hp, top speed limited to 80 mph (165 mph unlimited). It could accelerate from zero to 60 miles an hour in 8 s which was faster than a contemporary BMW 3 Series. 1117 units were produced 1996–1999 as a compliance vehicle for California. Its 6.6 kWh battery gave it a range of 100 km or 60 miles in moderate (probably ideal) conditions according to Car and Driver.[4] That gave the EV-1 an efficiency rating approaching 6 kWh/100 km which is astounding and most likely largely due to the cars very low aerodynamic drag of 0.19. For reference, the somewhat boxy BMW i3 and sleek Tesla Model 3 have Cds of 0.32 and 0.23 respectively (Fig. 9.5).

The more air a car pushes, the more energy it needs to travel down the road. Efficiency is a large issue for EVs clearly but there is another reason for wanting a low Cd in an electric car and that is noise. Pushing or accelerating more air creates buffeting and vortices which create noise. In an internal combustion car, noise levels are inherently higher whereas electric drive systems offer a much lower noise floor. Hence EV designers have to take much more care to reduce wind and tire noise so these sources don't become prominent.

Several levels up from a golf cart, the Th!nk was the author's first EV driving experience. It was able to make it up a 400 m col. in Oslo but at a significant price in battery capacity. However, on the way down, the regeneration

Fig. 9.5 EV-1 was a great effort by GM and should have been further developed

[4] https://www.caranddriver.com/reviews/a32944084/tested-1997-general-motors-ev1-proves-to-be-the-start-of-something-big/

Fig. 9.6 The Th!nk showed the future direction of the urban car

restored a satisfying amount of power. Crude? Definitely. Fun? Absolutely (Fig. 9.6).

The Th!nk City was an electric city car produced by Norwegian carmaker Think Global from 2008 to 2012. It was perhaps optimistically rated for a top speed of 110 km per hour (68 mph), and a range of 160 km(99 miles).

Hybrid Stepping Stones

Hybrids began to appear several years after the EV-1. These used electric motors and very small batteries to augment the main gasoline engine.

2000 Honda Insight

- "The Egg" was a small two seater with faired rear wheels, manual gearbox, a tiny battery and a 13 hp electric motor to help its 67 hp 3 cylinder internal combustion engine. It never ran entirely on electricity but shaving the peaks off demand and shutting down the ICE when not needed, allowed it to achieve over 60 mpg (US) in the right hands (Fig. 9.7).

Fig. 9.7   Honda jumped out to an early lead with its first hybrid, the Insight, but then faded

Toyota Prius

- Toyota equipped its 2000 Prius with the most advanced hybrid system of the day and one which has stayed class-leading for two decades. It allowed the ICE to shut off and run simply on its electric motor whenever the load was low enough. So driving at low speeds under light throttle or down hills at highway speeds, the car would use no gas. It would also, like every hybrid after the Insight, recharge its batteries when slowing down or braking. But still, all electrical energy that went into the small battery (0.9 kWh) came from either the gas engine or regeneration. There was no plug-in capability. The Prius had slick aerodynamics to improve efficiency.

Chevrolet Volt

- General Motors broke new hybrid ground when they launched the Chevy Volt in 2011 with a large 16 kWh battery and 149 hp electric motor which was strong enough to both run at highway speeds and cover 50 miles without having to call on its ICE for support. The Volt was the first true plug-in hybrid and its battery could be fully charged from a household 120 V outlet overnight. This meant that most people could run their errands during the day in solely electric mode and plug-in when they got home and repeat the next day without ever having to visit a gasoline station. If they did need to make a long trip, the ICE generator (it didn't drive the wheels but only recharged the batteries) would get them there, stopping only every 400 km for gasoline. This new PHEV (Plug-in Hybrid Electric Vehicle) template has been adopted and modified by many manufacturers in the past decade.

Mainstream Milestone EVs

Nissan Leaf

• Nissan led the world in producing the first truly mainstream EV, a car which could provide all of the normal functions and be driven in exactly the same way an ICE car could but one which ran only on electrons. In 2010 Nissan introduced the Leaf with its 24 kWh battery and perhaps a 70 mile or 110 km range on a good day. The 2022 Leaf has a 62 kWh battery offering almost triple the range of the original model and has a much more powerful motor. The Leaf brought everyday EV practicality to the world.

Tesla Model S

• Launched in 2012, the Tesla Model S hit the road as the first complete EV offering for consumers. The vehicle and the Tesla charging network constituted a very compelling package. The Model S landed in the middle of the most desirable sports sedans available at that time according to Car and Driver magazine. It found the Tesla tied for first in braking and tied the M-B AMG in acceleration and lateral grip. Although it came in around at $12,000 more than the Mercedes or the BMW M5, the Model S undercut the Porsche Panamera Turbo S by $70,000.
• By 2015 when its "Ludicrous" mode was installed it had become the fastest sedan on the planet and it also had the lowest luxury sports sedan running costs plus the smallest carbon footprint by far. The Plaid version of 2021 is the fastest production car in the world and was done, not to appeal to buyers, who will never use its violent 6 s to 200 kph acceleration, but to prove the point that EVs are simply faster than fossil fuel cars.
• The practical Leaf had been leap-frogged by what was widely seen as the most advanced automobile in the world.
• The high end Model S was followed up with the more mainstream Model 3 in 2017 continuing the company's technological lead over its slowly awakening competitors. But the legacy car companies were also beginning to produce very credible EVs (Fig. 9.8).

Fig. 9.8 A huge leap in both performance and technology, the Tesla Models S was a shock for established luxury car makers

BMW i3

- The BMW i3 did for the chassis and construction of electric vehicles what Tesla had done for the electronics. Its advanced carbon fibre body was both ultra-strong and ultra-light, countering the major disadvantage of EVs; battery weight. Despite the simplicity of EVs and reduction in the number of components, the battery typically adds 200–400 kg (450–900 pounds) to the weight of the car. The i3 has a smallish battery but, in 2022 has a range of 250 km and weighs 1500 kg, hundreds less than most EVs. And, of course, it is a delight to drive (Fig. 9.9).

Chevrolet Bolt

- GM's first ground-up, mainstream EV, is a very compact and solid package with sporty driving characteristics and a great driving position and visibility. The somewhat plasticy interior and thin seats were upgraded in the 2022 model and the range is now close to 400 km. It offers greater practicality than the Tesla Model 3 and is considerably less expensive.

Hyundai and Kia

- The Korean twins introduced their Soul, Ioniq, Kona (car reviewer favourite) and Niro EVs over the space of 3 years and are now onto their second generation with the striking Ioniq 5 and 6. Hyundai is well positioned to

Fig. 9.9 BMW i3 is very innovative and will hopefully be followed by other similarly constructed models

take large chunks of market share from the bafflingly EV-hesitant Toyota and Honda.

VW ID.3

ID.3 was the name chosen by VW for its first clean-sheet EV design because three designates this as the third pivotal model in Volkswagen's history. The first was the Beetle (20 million sold), the second was the Golf / Rabbit (35 million sold) and ID.3 is the model, or at least the platform, which VW expects will carry it forward for the next 3 or even 4 decades. That is quite a legacy to uphold but the ID.3 appears to be a refined, sporty, practical and efficient vehicle that will keep VW factories operating at capacity for many years (Fig. 9.10).

Ford Mustang Mach e

Ford arrived a little late to the party with its first ground-up EV but the new Mustang has received very favourable reviews for its build quality, sportiness and overall implementation. It compares most closely to the Tesla 3. The fact

Fig. 9.10   VW commits to an all-electric future with the ID.3

that Ford assigned its most revered name to their first EV indicates the transition to electric is well under way.

Most manufacturers have committed to basing their EVs on new "platforms" which can be enlarged or shrunk to serve as the base for future models. This signals a heavy commitment to the EV market.

The list of manufacturers producing EVs is a very long one. The above models stand out because they marked the progression of technology and market penetration of EVs. Most major manufacturers have now made very determined entries into the EV market. Honda and Toyota, the first hybrid pioneers, are notable laggards. Perhaps their early but misplaced bets on hydrogen delayed EV development but certainly they are beginning to show signs of EV revival and it is likely they will have full lineups of fossil fuel free models by late-decade.

Whatever the timetable, it is generally recognized that transitioning to virtually 100% electric car production will happen in the next 15 years. Many companies have announced target years for their withdrawal from the ICE market. Some countries have set dates by which all new vehicle sales must be battery-electric.

That Was Then, This Is Now

EVs are all about efficiency, efficiency, efficiency, particularly if you are an engineering firm like Mercedes Benz.

The EQS is the pure-electric Mercedes many have been waiting for. With a claimed range of 700 km on a single charge and 329 or 516 bhp, it is also the most aerodynamic production car ever made, with a drag co-efficient of just 0.20 (Figs. 9.11 and 9.12).

Author's Rant

Why, oh why, would anyone consider a gas or diesel engine vehicle which will be inferior in every aspect of performance and comfort? The EQS is so quiet Mercedes-Benz found it necessary to offer the driver the choice of dialing in different types of added noise. EVs now set the high bar for luxury, refinement, performance, efficiency and low long term cost.

Why would a person want to buy an ICE vehicle unless they have a really specific need that an EV simply can't fill?

The Mercedes-Benz EQS with batteries of up to 107 kWh is very fast, very quiet, very capable and tending towards very efficient. How does a large S-class Mercedes rate mention of the word "efficient" in its description? 700 km on 107 kWh yields a consumption of just 0.150 kWh per kilometer which would be respectable for a compact sedan.

The EQS will do to the luxury car market what the Tesla Model S did to the sports car market. It will signal that if it isn't electric, it may be good but it will never be the best.

Like cell phones, EVs now garner almost as much attention for their screens as they do for their range or performance. When is too much information simply just too much? (Fig. 9.13)

Fig. 9.11 Ferdinand Porsche' starting point shows how far EVs have come. Aerodynamics was a focus early on

Fig. 9.12   The M-B EQS proved that EVs can dominate any market segment

Fig.   9.13   M-B   EQS   dashboard.   Autogefuhl.   (www.youtube.com/
watch?v=WKrNCVSdefA)

The Market Shifts and Then Landslides

Porsche, BMW, Audi, Jaguar and Mercedes didn't get into the electric car market when Nissan proved the viability of EVs with the Leaf. These high end sports sedan manufacturers got into the market when start-up Tesla started eating their lunch by producing a vehicle that was faster, more refined and more advanced in many other ways than their ICE offerings.

The Model S and Model 3 demonstrated that not only were EVs practical but they were superior in a number of areas which previously had been the exclusive domain of these prestige manufacturers. Also, in terms of cold hard accounting and survival, the Tesla Model 3 began outselling all of their comparable models combined in some markets. The major manufacturers didn't need the motivation of romance or public displays of virtue to launch themselves into action.

When Teslas of all stripes began beating the luxury performance brands on the drag strip and on the track, the traditional manufacturers had to respond or be reduced to irrelevance. Who won the 1948 Lemans race mattered little if the cars in their current lineup were being blown off decisively by new technology offerings, ones which, by the way, were cheaper to run, more reliable and almost certainly the way of the future.

Automotive executives can read not only sales charts, they can also absorb climate graphs and oil reserve reports. Clearly, the business environment for fossil fuel vehicles was going to become far less rosy and the need to transition to much cleaner and efficient vehicles became a corporate imperative (Fig. 9.14).

What Does the Above Chart Say to an Automotive Executive?

In Norway, where EVs currently have over 50% of the new car market, it is becoming more difficult to sell a used gasoline car and prices for them are falling. Norway will ban the sale of new gas or diesel or hybrid vehicles by 2025

Fig. 9.14 Sports sedan sales chart as Model 3 rockets to number 1

and Britain by 2030. Ford and Jaguar have announced that they will only produce 100% electric vehicles by 2030.

Around 6.5 million electric vehicles were sold worldwide in 2021. Sales are expected to keep on growing exponentially going forward.

Attitudes are changing and costs are becoming much more clear. Polls show the following factors reducing customer reluctance:

- Increasing difficulty of selling used ICE cars in some markets / depreciation
- Confidence in batteries lasting the life of the car
- Stigma of carbon emitting cars
- Fear of buying retro-tech
- Teslas have the highest owner approval ratings across all demographics

EV Market Overview; the Empires Strike Back

Who isn't in the EV business now?? Tesla rose from 0 to a dominant market force in 10 years. Amazon, Facebook, Dyson, Sony and Microsoft have invested in the EV car business. Basically anyone, from internet to vacuum cleaner companies, with a few billion to burn apparently sees barriers to entry tumbling. This is due to the lower investment in drivetrains/engines tooling and production and certification being vastly lower for EVs than for ICE vehicles.

Engineering can be rapidly acquired from auto parts manufacturers such as Magna. Will the automotive industry change to the point where suppliers produce all of the building blocks for vehicles and name brand manufacturers merely customize them into the final product?

Established, "legacy" manufacturers such as M-B, Ford, GM, VW and Jaguar have made decisive commitments to EVs. But China is coming on as the preeminent industrial power and is now challenging to become the dominant technology power.

Still, western producers are now following the Tesla model by building their own batteries and charging networks. As an example, Volkswagen plans to build six "gigafactories" in Europe by 2030[5] providing enough capacity for four million Volkswagen ID.3 s as well as adding a huge number of charging stations across Europe, the United States and China. It plans to recycle up to 95% of the raw materials involved in battery production as the majors begin to deal with all aspects of vehicle production and how their vehicles fit into broader energy strategies (Fig. 9.15).

---

[5] https://www.cnn.com/2021/03/15/business/volkswagen-batteries-electric-cars/index.html

Number of models available in key
regional markets

Fig. 9.15 China takes the lead in car manufacturing

According to VW Chairman Herbert Diess, Volkswagen's vehicles will connect to private and public energy systems saying "Electric vehicles will become mobile power banks."

A 100% electric future? Porsche's head of global sales, Detlev von Platen commented "That is the future, full stop."

GM is ramping up and there will be many such announcements in the coming years as the societal transition from fossil fuels to electricity "gathers steam". Maybe our grandchildren will use the phrase "builds charge".

Why Not Hydrogen?

This can be a lengthy topic but the appeal of hydrogen is largely found in its fast re-fueling time, very comparable to gasoline, and somewhat lower weight than batteries. Beyond that, hydrogen negatives include:

- It's a very lossy process with a lifecycle efficiency in the range of 30%
- It is hard to contain
- Requires heavy tanks rated for ~10,000 psi
- And it is corrosive

Hydrogen definitely has its applications but it will probably never become a mainstream energy storage medium for the ground vehicle fleet.[6]

EVs on a Mission

As the number of EVs surges, new models are now being aimed at every conceivable niche of the market. Small, cool running and quiet electric motors allow designers a much broader canvas on which to express their innovative talents. For marketing managers this translates to "if you have a special need, we will fill it or, if you don't, we will invent one for you".

In 2022, EVs are taking many forms.

City EVs

Designed for mostly in-town or between small town use, they will work very well on the highway but require more time to make long journeys. These vehicles are small on the outside but have useful cargo capacity and very tight turning circles.

In order to get smaller packages and lower prices, battery size, interior space and sometimes interior finish were sacrificed. In some models though, like the Honda e and Mazda 30, interiors have been made distinctly upscale. But these cost more.

Current models in this segment include the Honda e, Mini e, Smart Ed and the Renault Twizy which offers the most fun you can have in a car when it isn't raining or snowing (Fig. 9.16).

---

[6] Hydrogen for cars vs battery and ICE efficiency comparison - Just have a think, https://www.youtube.com/watch?v=IEdPp7_lJes

Fig. 9.16 True micro EV - Renault Twizy. Real wind-in-your-hair motoring because there are no windows

All-Round EVs

The Chevy Bolt, Nissan Leaf, VW ID.3 and Hyundai Kona and Ioniq are representative of these sedans and hatchbacks which will cover most people's needs most of time while being enjoyable and even sporty to drive.

Performance

What is meant by "performance"? How about 0–100 km times in 5 s? This once rarified air can now be penetrated by almost pedestrian EVs and is a mark most EVs with sporting pretentions can exceed. From mainstream entries like the Tesla Model 3 to the Jaguar i-Pace and Porsche Taycan with its 750 hp, there are dozens of entries offering prodigious torque, low centres of gravity and instantaneous response which are design features essentially baked into every EV.

For driving enjoyment, judge for yourself from a comment from Graham Fletcher, an auto-journalist at driving.ca: "The resulting rush will leave your stomach at the starting point as the Taycan Turbo S warps forward; it's quite simply the fastest car off-the-line I've ever tested. Of course, the 1.7 seconds it takes to complete the 80–120 km/h passing move also says it all." Graham also found the Taycan handled exceptionally well. But why call it a "Turbo" if it doesn't have a turbocharger?? Old marketing references die hard.

Models from every Tesla to Lotus to converted Ferraris fill out the performance category.

But the ground-breaking point is that all of the best new models, regardless of category are electric. The M-B EQS is probably the most advanced luxury car in the world and the Porsche Taycan Turbo S, takes the crown in the GT Touring and Sports category.[7] The most refined small cars are electric and now EVs are not simply the practical choice but the aspirational one as well. Across the board, EVs have raised the bar for efficiency, performance, luxury and refinement.

Utility: Genuine Utility

EV design has emphasized efficiency as the low aerodynamic drag numbers of hybrids and more advanced EVs demonstrate. In order to reduce our footprint it is critical to both transition to electricity and reduce our demand. That means smaller vehicles with smaller batteries. But many of the designs simply electrify SUVs and trucks and still play to the wretched-excess gene with their extremely high weights and barn door-like Cds.

Do vehicles have to be urban assault capable or can they just be comfortable, practical, and easy to handle transportation? (Figs. 9.17 and 9.18)

Compare the ultra-high efficiency Aptera with its 56 kWh battery and 1000 km range to the Hummer EV and its mammoth 200 kWh battery which will deliver 560 km of range.

No the Aptera[8] can't tow a cruise ship but most of the time these two cars would be carrying the same load which is one or two passengers. The Aptera can go 3 times further than the average EV on same size battery courtesy of its small size and Cd of 0.15. It can go over 6 times further on the same energy

---

[7] https://www.youtube.com/watch?v=7qyKhP4lzsY
[8] https://www.aptera.us/

Fig. 9.17   Aptera. The efficiency leader

Fig. 9.18   Electric Hummer

as the 9000 pound (4100 kg) Hummer EV. Sustainability requires that we electrify and downsize, rather than electrify and supersize.

Although they are the last segment to go electric, eTrucks will arrive in abundance by 2023 with the major manufacturers all offering battery versions of their most popular models. And, as this book is published, the Ford Lightning is hitting public roads dead in the centre of the last large consumer market yet to have major EV penetration. The Lightning is as much of a tipping point as the Tesla Model 3 and it represents the beginning of the end of the era of large fossil fuel engines in consumer vehicles.

Ultra-Performance Is Now the Domain of EVs

Lotus cars have traditionally blended performance, handling and efficiency into very sporty packages. Their Evija model, enabled by the inherent advantages of electric motor and batteries, takes this performance formula to the extreme.

The Lotus has 2000 horse power in four 500 hp motors, one for each wheel, a 300 km plus range and an efficiency level to rival an EV compact like the Kona, (when driven like a daily runabout) (Fig. 9.19).

With the packaging advantages and high power performance of electric motors and batteries, expect a continuous stream of hyper-cars from Europe, the USA and China and, given the low entry costs, many points in between.

Compare the Evija to the muscle car Dodge Hellcat. The 2000 hp of the Lotus propels it from a standing start to 100 km per hour in under 2 s with a top speed of over 300 kph. It uses the energy equivalent of 2.7 litres of gas to cover 100 km while the Hellcat uses 14 litres or 5 times as much.

The Hellcat is slower, noisier, doesn't handle as well, has only two wheel drive and vastly higher energy consumption. Its GHG emissions are scores of times higher. It is currently a lot cheaper but given its noise, pollution levels and retro handicaps, it is really what young boys now dream of owning? (Fig. 9.20)

Fig. 9.19   Lotus Evija - supreme performance looks like now

Fig. 9.20   Dodge Hellcat - what muscle cars used to look like

Future Market Trends: Will EVs Become Cheaper?

Do you indulge in stock picking car companies? If they are traditional ICE manufacturers, how well are they managing with their stranded asset strategy? As they transition to EV manufacturing, how able are they to disentangle themselves from ICE engine and component manufacturing without suffering massive write-downs?

Wright's Law.[9]

While studying airplane manufacturing in the 1930s, T.P. Wright determined that for every doubling of airplane production the labor requirement was reduced by 10–15%. In 1936, he detailed his full findings in the paper "Factors Affecting the Costs of Airplanes." Now known as "Wright's Law", or experience curve effects, the paper described that "we learn by doing" and that the cost of each unit produced decreases as a function of the cumulative number of units produced.

Therefore EV prices will drop faster than ICE prices because ICE production, at 90 million annually, will never double again and EV production is doubling every 2 years so that every 2 years, prices should drop 15% barring a battery breakthrough in which case it would be more. Note that this trend could be thrown off by material shortages.

The continued shift of both industrial productive capacity and technological expertise can be seen in China's booming EV sector. Is China now leading

---

[9] https://www.youtube.com/watch?v=7lKRL7Hghfk

the world?[10] They are introducing 80 new EV models in 2021 ranging from cheap to cool to crazy. Costs are a function of exchange rates which are likely rigged to a greater or lesser degree[11] but the vehicles China is producing are real and so is their increasing share of the international vehicle market.

As EVs increase their share of the market, the market dominance of China is likely to increase. Manufacturers in Europe are adjusting quickly but those in North America and Japan are lagging badly. Tesla excepted of course.

Northern Comfort

The graph on the right disposes of the issue of cold weather performance. Advanced northern countries are rapidly adopting them. Are EVs superior to ICE cars in extreme cold? No. Can people (mostly) adapt easily to them? Yes. A heat pump can make a large difference in low temperature performance and range.

Here is a quote from a snow country resident in Central Ontario, Canada. "Hyundai Kona upgraded the software which improved range by 70 km. With remote defrosting I only have to scrape off the car 2 or 3 times a year. Wife loves it. Warmup on grid power which means 100% of the energy is applied to heating compared to a gas vehicle where one left idling to heat it is probably less than 10% efficient. Electric means near instant heat and defrosting - heat by the end of the block" (Fig. 9.21).

Picking the Right EV

Which is your market category? This substantially influences the price ratio of EVs to petrol cars.

- EV pricing is much more expensive for basic cars
- EV pricing is pretty close for luxury cars
- EV pricing is very attractive for sports sedans e.g. Tesla 3 vs BMW 3 Series

The trend seems to be that the more expensive, higher performing and luxurious the car, the less the cost differential is. Possibly the inherent attributes of

---

[10] https://www.youtube.com/watch?v=rCNjDTiq4Rc

[11] Certainly lower prices are guaranteed by lower or non-existent labour and environmental standards compared to the western democracies.

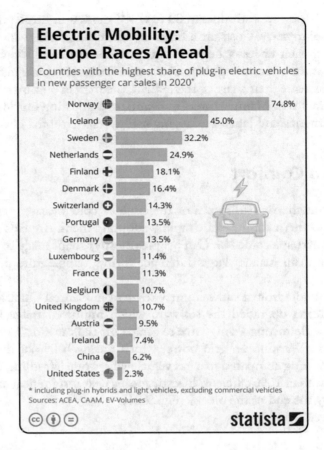

Electric Mobility: Europe Races Ahead

Countries with the highest share of plug-in electric vehicles in new passenger car sales in 2020*

Norway	74.8%
Iceland	45.0%
Sweden	32.2%
Netherlands	24.9%
Finland	18.1%
Denmark	16.4%
Switzerland	14.3%
Portugal	13.5%
Germany	13.5%
Luxembourg	11.4%
France	11.3%
Belgium	10.7%
United Kingdom	10.7%
Austria	9.5%
Ireland	7.4%
China	6.2%
United States	2.3%

* including plug-in hybrids and light vehicles, excluding commercial vehicles
Sources: ACEA, CAAM, EV-Volumes

statista

Fig. 9.21 EV market penetration

an electric drivetrain reduce the development and production costs of higher end vehicles and these substantially offset the cost of the batteries.

When test driving EVs, individuals might be struck by their homogeneity because of the difficulty in telling the difference between their motors and transmissions. The only difference really is just that one engine has more power than another whereas internal combustion engines and their transmissions vary tremendously in their sound, feel and responsiveness.

The boundaries of the car categories are becoming smeared and transformed as new players with revolutionary products stomp on the toes of existing market conventions. The Tesla Model Y beats the latest model of classic sports car heavyweight, Chevrolet Corvette to 100 km/h and comes close to its time around a track. It hauls your groceries, 7 people and requires approximately 13% of the energy the Corvette uses.[12] Carbon footprint? Do you have to ask?

EVs make it easy to go green because it is like having your cake and eating it too. With less maintenance, lower cost, better performance and lower Noise Vibration and Harshness (NHV) an EV is now not just an easy choice but also, probably the logical one.

And there is a lot of enthusiastic help available to ease your entry into the EV world. Here is the thought process and experience of one young buyer in the UK, documented in a series of Fully Charged videos. "Maddi Goes Electric" videos[13] are worth watching if only for the positive energy they exude.

There are NGO and commercial organizations to help with decision making and, of course, a nearly unlimited amount of YouTube reviews from professionals and amateurs alike.

Organizations such as Plug and Drive Canada offers "Find Your EV Match"[14]

- customizable search filters that can be used to set unique daily driving distances, the duration of leasing or financing agreements, upfront down payment amounts and model year/brand preferences
- Total cost of ownership calculator customized by where you live simply by entering your postal code
- A side-by-side comparison tool that shows the difference between an electric car and an equivalent gas car
- A database of EV incentives offered in Canada
- The Fully Charged Network covers the transition to a fossil fuel free society from soup to nuts.

There are all manner of guides from these groups such as the Green Buyers Guide[15] or EPA Green Guide[16] in the US to the Green Car Guide in the UK.[17] Despite the hurricane of information, just keep in mind these cars are less complicated than the ones you are used to.

In the rapidly evolving EV market, companies are adopting new design strategies. Rather than an all-out quest for range, some companies are making EVs either with much smaller batteries than normal or a range of batteries to

---

[12] At 0.16 kWh/km or 16 kWh per 100 km vs 12 L/100 km. (12 L gas × 10 kWh energy content/litre = 120 kWh/100 km

[13] https://www.youtube.com/watch?v=9A4ytljB-jo

[14] https://ev.plugndrive.ca/

[15] https://www.greencarreports.com/news/buying-guides

[16] https://www.epa.gov/greenvehicles

[17] https://www.greencarguide.co.uk/

lower the initial cost and / or putting more into interior touches. As EVs evolve and battery prices drop, it is no longer just about range. Also, consumers have a much better understanding of their needs and no longer have to be paralyzed by range concerns as the number of rapid chargers increases rapidly.

Strategically, governments need to get involved to completely build out charger infrastructure and encourage or standardize small batteries to take the pressure off material supplies.

Chevrolet was shooting for maximum battery size in the 2017 Bolt while BMW and Mazda opted for a more sophisticated product and used battery funds to pay for it in their i3 and MX-30 models. Some manufacturers are now reducing the battery size to get the lowest price and highest efficiency as Hyundai did with its Ioniq.

Of course not everything can be measured in direct dollar cost, performance, range and amenities. Governments have offered a wide range of incentives to encourage EV adoption including:

- Free parking
- Lower road tax
- No charge for entering city centre
- Access to green vehicle lanes
- Free electricity
- Lower insurance
- Accelerated income tax expense write-offs

Worried about making a satisfying purchase? Don't be. EVs are the most satisfying cars to own according to Consumer Reports with the first choice of the Tesla 3 by 4 generations; millennials, gen x, boomers and the silent generation.

Not certain what you should do? Maybe the best choice is to keep your current ICE vehicle for another 1 or 2 years during which time EV costs will have gone down while choices will have increased. It is estimated that by 2025, there will be 400 different models of EV to choose from. Of course, as time ticks by the trade-in value of your gas guzzler may well decline at an ever-increasing rate.

The next vehicle you buy will likely have a 15–20 year life so make sure you aren't investing in assets which will become stranded a few years down the road.

Note that many experts predict the purchase price of EVs will go below that of ICE vehicles in 4 or 5 years. That is hard to imagine because EVs are heavier and use materials generally more expensive than the main drivetrain component of fossil fuel cars which is steel. Price parity may be approached however and certainly long term costs will be significantly lower.

Used EVs

Buying a used EV is becoming a popular choice among those faced with the necessity of buying another car and can't commit to a new EV yet feel they need to move on from petrol vehicles. Some used car businesses are focusing strictly on EVs and many offer much more than a typical used car lot or even conventional new car dealership.

"All EV"[18] is a chain of used EV car stores in the Canadian provinces of Prince Edward Island and Nova Scotia. They offer EV educational courses on vehicles and all government rebates and grants plus charging installation info. Each vehicle comes with a battery health report.

One example is a 3 year old 2018 Tesla 3 with 92,000 km and, only 6% battery capacity loss indicating a thoughtful prior owner. One 8 year old Tesla S with 250,000 km had only 10% degradation, a testament to both the owner and good battery management software.

Batteries are turning out to be much more long-lived than anticipated when the first EVs were produced 11 or 12 years ago but their condition is still one thing a prospective buyer must get professional help with. Early Leaf batteries were not liquid cooled and therefore more temperature sensitive than later versions. In hotter climates they suffered more rapid degradation than in cooler climates.

The whole used EV / used gas car equation is changing rapidly as large numbers of EVs are now coming off lease. Search "buying used EVs" and only go to pages published in the past year. But keep in mind that even though the electric motor should last for decades and the brakes and drivetrain should also need far less upkeep, EVs shares body, interior, steering and suspension parts with their ICE forbearers. These will likely be no more or less durable than in a "conventional" car.

The EV Charging and Range Equation

When it comes to charging networks, there is Tesla and then there is everyone else. In the Tesla system, the driver plugs the charging cable into the car and charging starts immediately. The system recognizes the car and will charge the owner's account if they don't qualify for free charging. It is quick, efficient and simple but is restricted to Tesla's vehicles although this may change.

---

[18] https://www.allev.ca/

Other charging networks can approach this simplicity but many have processes far more complicated than the existing model of gasoline fill-ups. Instead of swiping a credit or debit card and then pumping energy into your tank, some networks force you to have registered with them and use their phone app to begin buying energy.

Beyond that, some of these systems are not glitch free, leading to a long and frustrating process with some numbers of people abandoning the effort and looking elsewhere for a charge. Clearly, it is early days in this area but the existing gas pump model surely can't be difficult to emulate just because the fuel is electrons instead of liquid carbon. Despite the current variability, expect this industry to learn as rapidly as the rest of the EV industry chain.

Do's and Don'ts of Charging

Do

- Charge at low rates at home overnight whenever possible
- Allow the battery to pre-condition itself, i.e. heat or cool itself to the ideal temperature for both use and charging
- Charge before capacity drops below 20%
- Charge to only 80% mostly

Don't

- Charge at high rates frequently
- Charge to 100% repeatedly
- Use battery to very low capacity repeatedly
- Charge battery when it is very hot or very cold

Charging in USA: "Almost Tesla-like" It seems Plug and Charge is bringing Tesla-style convenience to non-Tesla owners. The Plug and Charge system determines what kind of EV you have and simply bills you. According to one owner "Lo and behold: I watched the machine quickly identify the car, validate the charge, and start the current flowing. No fob, no app, no toll-free number to call." Several other networks use similar protocols.[19] And then

---

[19] www.youtube.com/watch?v=0zOoJh5BFx0

Witricity

Fig. 9.22   Induction charging

there is the ultimate in convenience which is wireless EV charging that occurs when you park your car over a charging pad (Fig. 9.22).

There are different types of public chargers ranging from 120 V household current to 800 volt DC systems. Induction charging gets rid of the driver involvement except hit a button on their way out of the car. This is claimed to be as efficient as plug-in charging but precise positioning of the car over the induction loop is critical. To accomplish this, the self-driving feature of the car can be employed (Fig. 9.23).

Suffering from charging confusion? Let Nicki Shields of Electrifying[20] sort things out for you. Or you can buy an EV and become an expert in a very short period of time. It just isn't that difficult.

One way of looking at the rate of charging is by equating power to range so that several minutes of charge at a certain rate will produce so many kilometers of travel distance. This is useful as there will be few occasions which require a driver to charge from zero to 100%. Charging strategy will likely relate more often to how far one expects to travel than to the capacity of the battery pack (Table 9.1).

A large caveat should be mentioned here and that is charging rates are not a straight line affair. The rate of charge for most vehicles is greatest between 10% and 80% and slower at the margins and during extreme weather. Audis

---

[20] www.youtube.com/watch?v=0zOoJh5BFx0

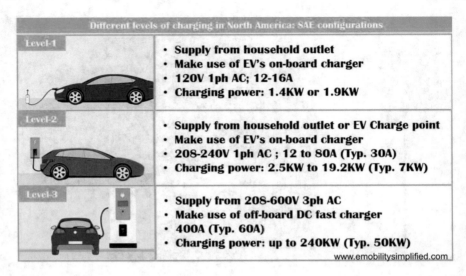

Fig. 9.23 Levels of charging systems

Table 9.1 Charging times by charger type

	Energy consumption of car in kWh/100 km	Time to charge for 100 km			
		120 V house charge 1.5 kWh	240 V 3 kWh	40 kWh	150 kWh
Hyundai Ioniq	13	9 h	4.5 h	20 min	5 min
Tesla 3 standard	16	11 h	5.5 h	24 min	6 min
Jaguar I-pace	25	17 h	8.5 h	38 min	10 min

appear to be able to accept high rates of charge even to near 100% but with most vehicles, it is better to make a few short stops when the battery is low than to try and fill up completely which will take longer. And, of course, the objective for most people will be to only use charging enough to get back home where the electricity costs will be far lower.

The ability to "zap and dash" is dependent on the maximum rate of charge a car can maintain. This is a very dynamic area of development with some manufacturers now offering vehicles with smaller batteries to keep costs and weight down but compensating for this partially by providing very high rates of charge. Not only will the mileage and charging times vary, so will the costs.[21]

---

[21] https://greencarjournal.com/electric-cars/what-does-public-charging-cost/

Now, in many new models, charging can go both ways. V2G (Vehicle to Grid) or V2L (Vehicle to Load) capability means that you can pretty well operate any household appliance from your car as well as charge other EVs at a low rate. During power outages, it will genuinely pay to have a rolling battery in your driveway.

Cost of Charging

Novice inquiry: Where do you charge your car? Reply: Where do you charge your phone? Bingo! The vast majority of people will charge their EVs overnight at home when rates are the lowest.

In a real world analysis of his own charging experience Andy Sly[22] details the cost of actual electricity use for his Tesla S (high performance sedan) over 75,000 miles. It came out to 18,600 kWh used for 250 W/mile which equates to 15.6 kWh/100 km. Note that Andy lives in a moderate climate and may not even own snow tires.

The plug-to-wheels efficiency = 94% so the actual energy used to charge the car was around 20,000 kWh. Andy spent $1350 US on home charging so far and 92% of his charging was at home. At 8 cents /kWh (exceptionally low rate) in Kentucky the cost worked out to 1.8 cents US per mile or 1.1 cents per km.

Compare this to the cost of fueling a BMW 3 Series which averages 28 mpg producing an total gas cost of $9300 or 12.4 cents per mile, over 6 times as much as the EV.

And yes, new EV owners charging from home will probably want to install a charger with 4 or 6 times the capacity of a 120 V outdoor wall plug and this will cost in the neighbourhood of $1000. These should last decades.

An individual's cost of charging at home will almost always be the lowest cost option as electricity rates are typically lowest overnight. It makes sense to only use charging stations for partial charges to last long enough to allow the driver to return home.

In the US, residential average per kilowatt-hour rates currently range from 9.3 cents in Louisiana to 28.9 cents in Hawaii with the overall national average being 13.3 cents. California, has a residential average cost per kilowatt hour is 20.1 cents. The EIA state-by-state list of rates is here.[23] Look on your

[22] https://www.youtube.com/watch?v=FRie5t9GHLw
[23] https://www.eia.gov/electricity/monthly/epm_table_grapher.php?t=epmt_5_06_b

monthly hydro invoice to see exactly what rate you will be paying to "fill-up" at your own house.

Rates are around 20 cents per kWh in the UK, 35 cents in Germany and 10 cents in China. Coal is the low cost fuel for electricity generation but as countries go green, rates will rise from their carbon fueled lows. Rates will also become much more variable so the ability to time your charging events will have a large effect on driving costs. Edmunds reviews the cost of EV charging in this footnote link.[24]

Speed, Tires and Weight

Aside from aerodynamics, the rolling resistance of tires and the weight of the vehicle determine how far an EV can travel on a given amount of power. The choice of putting larger wheels on a specific car means they will have a lower profile (shorter sidewalls) which provides better handling but higher rolling resistance which decreases range and efficiency. Also, low profile tires are less forgiving over surface irregularities and produce a harsher, noisier ride (Fig. 9.24).[25]

Above are the results of a scientific study of speed and tire size on range.

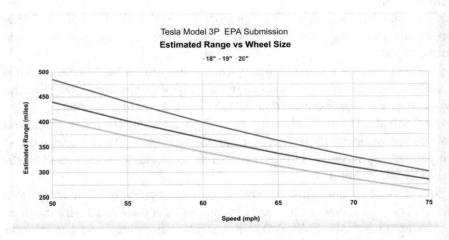

Fig. 9.24   Tires and speed matter for efficiency

---

[24] https://www.edmunds.com/fuel-economy/the-true-cost-of-powering-an-electric-car.html
[25] Geek presentation of math - https://www.youtube.com/watch?v=NYvKxsYFqO8

On the internet, there are large numbers of amateur and scientific real-world testing of EV range under different conditions. Here an amateur German enthusiast looks at the question on his YouTube channel "Battery Life". He found the VW ID.3 got 440 km of range when cruising at a steady 90 km per hour on the autobahn but only 280 km of range at 130 km/h. At 90 kph, the VW used 0.13 kWh per km and at 130 kph used 0.2 kWh/km, a 50% increase in consumption.

Bjorn Nyland[26] has tested numerous EVs over large distances in harsh conditions while some reviewers test in convoys.[27]

Below is a sampling of the virtual Niagara Falls of data available to those researching EV efficiency and range (Table 9.2).

The above chart is from the "Battery Life"[28] YouTube channel, an amateur site dedicated to real life experiences of living with EVs. While not professionally polished it is honest and pretty representative. The same can be said of

Table 9.2 Speed vs efficiency for different cars by Battery Life

Consumption in Watt hours/ km							
	90 km/h	110 km/h	130 km/h	Month	Temp (°C)	Weather	Notes
VW eUp 2020	128	165	187	March 2020	7	Dry	Winter tires, not gps
Hyundai Ioniq 38 kWh	104	134	163	May 2020	23	Dry	
Renault Zoe ZE50	122	158	202	June 2020	24	Dry	AC
VW eGolf	116	147	192	August 2020	30	Dry	AC
Mini Cooper SE	117	152	204	September 2020	17	Dry	
VW ID.3	133	170	218	October 2020	12	Dry	19 inch tires
Nissan Leaf e+	159	189	234	October 2020	12	Dry	Winter tires
BMW i3s	149	186	247	March 2021	1	Dry	Winter tires

[26] https://www.youtube.com/watch?v=wF2Ub315Y60 (Note: the comparative charts are towards the end.)

[27] https://www.youtube.com/watch?v=t0KODBGQVCM

[28] www.youtube.com/watch?v=6Jo3IuskzD4

Bjorn Nyland's channel which features a great deal of EV efficiency information. Undoubtedly, there are test results available from individuals and perhaps agencies in your region.

As with any device and any fuel, moderate use will yield the longest ranges and hard or sporty use will require more frequent charges or "fill-ups".

Vehicle Life Span

The average car made in the 2000s will likely last close to 300,000 km. After that, the cost of maintenance becomes prohibitive. Car life varies by country and income level but steel bodies, suspension and brakes will last only so long. Battery life is now approaching the maximum average life expected of conventional cars.

Average lifespan

- Japan 12 years
- Canada 13 years
- USA 15 years
- Australia 10 years

The BMW i3 with its carbon fibre and aluminum chassis may be different. If the body lasts decades, maybe replacing the battery 15 or 20 years on makes sense. It certainly would be the low environmental impact choice and much longer consumer goods life is critical to reducing consumption.

"And, as their technology is perfected, electric cars have the potential to run longer. "Three hundred thousand miles (450,000 km) could be the standard for an electric car," says Laura Trotta. The engines have fewer moving parts, which reduces breakdowns, resulting in less maintenance and longer lifespans (the same is true for hybrid vehicles). Also note that EVs use battery regeneration when slowing down which dramatically reduces wear and tear on rotors and pads.

In July 2018, Tesloop, a Tesla taxi company, announced that one of its Model S cars had passed the 400,000-mile mark, and the company says it expects the car to last another 600,000 miles."[29, 30]

---

[29] www.consumerreports.org/car-repair-maintenance/pay-less-for-vehicle-maintenance-with-an-ev/
[30] www.aarp.org/auto/trends-lifestyle/info-2018/how-long-do-cars-last.html

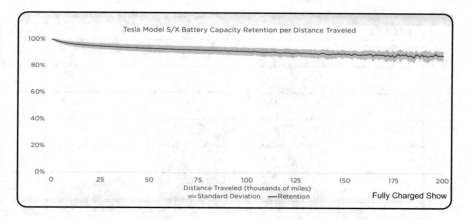

Fig. 9.25 Tesla multi-year report indicates very manageable levels of battery decline

Battery Life

Tesla's 2019 Impact Report indicated batteries in the Model S and Model X are lasting much longer than 200,000 miles or ~320,000 km which is the average ICE vehicles lifespan (Fig. 9.25).

Tesla batteries are now showing a capacity decline of approximately 10% after driving 200,000 (320,000 km). Tesla has exceptionally good battery management while early Nissans did not. Most manufacturers are likely now approaching the Tesla's sophistication level.

What does this mean? Unless you have a carbon fibre chassis as the BMW i3 does, the battery in your EV will likely outlive the car itself.

Dollar Cost of EVs

Electric vehicles are already substantially cheaper to run than their ICE counterparts but despite predictions of price parity by 2021, they still, in 2022, cost more to buy upfront. However, the gap is closing and may be eliminated in several years. At that point, EVs will be cheaper to run and maintain and will likely hold their value vastly better than the fossil fuel vehicles they are replacing.

What will happen to the depreciation curves of ICE cars as EVs take over is unclear but it would be unwise to bet on the continued shallow decline of the trade-in value of gasoline vehicles (Figs. 9.26 and 9.27).

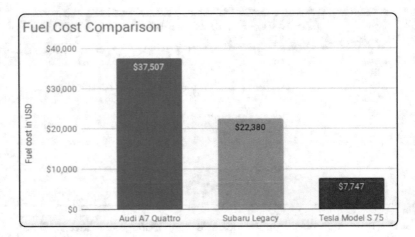

Fig. 9.26 Tesla ownership energy costs

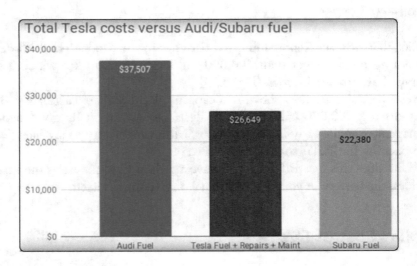

Fig. 9.27 Tesla total ownership costs

One EV owner, Sean Mitchell, presented a thorough and down-to-earth cost analysis in his video, "Do I recommend Tesla after 227K miles (365K km)?"[31]

Above is Mitchell's fuel cost comparison of two gas vehicles versus a Tesla Model S. Below is the total ownership cost of the Tesla versus just the fuel costs for the Audi and Subaru.

---

[31] https://www.youtube.com/watch?v=_OBxDDa8J14

Many amateur as well as professional reviewers[32] are publishing their cost estimates on social media and not all of these include a full set of numbers. EV enthusiasts are indeed enthusiastic and sometimes this overwhelms sound accounting practices. However, it is fairly clear that the long-term costs of owning an EV are significantly lower than those for an internal combustion engine vehicle.

Your Life and an EV

If you can recharge at home, don't go on many long (500 km +) trips, don't live in a region which regularly experiences temperatures of under minus 30 °C, basically the 2021 crop of EVs will work well for you. In terms of driver time, the trade-off is making no stops at gas stations and vastly fewer stops for maintenance and repair but plugging in fairly often at home instead. When the use of a commercial charger is required, a 100% charge may take 5–10 times longer than a typical gas fill-up.

By 2024, given the rapid development taking place, virtually all of the above cautions can likely be dropped. Apartment and condo dwellers may wait to see charging networks upgraded for their buildings so local access to fast chargers will be much more important.

This section is short because, in 2022, virtually anything a gasoline car can do, a battery electric car can do better and will cost less doing it.

Green Cred

Determining the overall impact of anything we produce and consume involves acquiring a lot of data. Once the conclusions fall out of the data analysis exercise, it is clear that EVs use less energy than ICE vehicles to move down the road and they produce lower GHG emissions. But what is upstream in the energy supply chain is also important.

Electricity solely from hydro, wind or solar sources allows an EV to discharge very low greenhouse gas emissions. If the source includes portions of nuclear, natural gas, coal or oil, then ghg emissions rise accordingly. Electricity in most countries is produced by a mix of sources with coal and oil fired generators being the dirtiest by far. But whatever the source, EVs still produce

---

[32] https://www.youtube.com/watch?v=KlKKUUmxlIw

lower emissions than fossil fuel cars, and in most cases, produce just a small fraction.

Emissions from natural gas plants are about 50% lower than those for coal and oil. Nuclear is very clean at the plant but it takes a good deal of fossil fuel to mine and process the uranium fuel. But coal and the dirtiest sources are being pushed out of the grid system by natural gas, solar and wind so emissions will drop on a continuous basis in virtually all regions of the world.

Charging your own vehicle from solar panels on your own roof is the cleanest way to power your transportation as well as the cheapest in the long run.

Because of their large battery packs EVs weigh more than internal combustion engine cars and consequently require more mineral extraction. This additional material content can be offset in the medium term by an increasing level of recycling. The carbon lifecycle of EVs and batteries is a well-covered technical subject.[33]

Consequently, in overall terms, EVs remain the way forward as illustrated by the Automotive Science Group presenting Tesla with ASG's Best Environmental Performance Brand[34] for the second consecutive year (Fig. 9.28).

Volkswagen feels it can achieve a 95% recycling level with its batteries and the major car producers also have a clear eye on designing recyclability into their products. But the best way to reduce material consumption is to simply either use the product for longer or re-purpose it.

Take for example the battery for a poorly treated 2011 Nissan Leaf. Imagine it now has only 50% of its initial 22 kWh capacity. Should it be thrown out and recycled or can it provide longer service in another application?

If the battery can hold a usable charge of 11 kWh, then it is almost matches a Tesla PowerWall at 13 kWh which costs close to $10,000. Placing the battery in a stationary commercial or residential application, where both temperatures and demand are moderate might allow it to provide another decade of work. Possibly even more as battery service, upgrades and control software become more sophisticated and cheaper.

There is an environmental cost to everything and the strategy to minimize these costs for EVs is to use them less, use them longer and don't use them at all if you can walk or bike to do what you need to do. The world won't be

---

[33] Effects of battery manufacturing on electric vehicle life-cycle greenhouse gas emissions
    https://theicct.org/publications/EV-battery-manufacturing-emissions
[34] https://www.automotivescience.com/pages/2019-study

Fig. 9.28 EV running off a dirty grid

saved by electrifying the large SUV fleet due to their weight and their 2–4 times greater energy consumption.

EV Highs and Lows

- Highs

  - The heat comes on "by the end of the block"
  - "I feel as though I've bought a luxury car."
  - "Yes, I do feel smugly superior to regular car drivers."
  - "I always felt stupid sitting at a stop light or going down a big hill and the engine was still running. Now the motor is either off or it is in regen!"
  - "Throttle response and one pedal driving – I'm never going back to gas."
  - "I haven't visited a gas station since I bought this thing."
  - "I can run the cottage off my Nissan Leaf."
  - It is possible to have performance, quietness, smoothness and efficiency.

- Lows

  - "Why can't they make charge stations as easy to use as gas pumps? I mean WHY!"
  - "We had an unexpected trip come up and had to cover several thousand kilometers in a couple of days so had to rent a regular car."
  - "I've surprised a few people who didn't hear me coming."
  - "The tires on my BMW i3 don't last longer than 20,000 km."
  - "Electric cars complicate the work of firefighters."
  - "The biggest battery option does deliver greater range but decreases both performance and efficiency due to increased weight."

Buy Now, Save Later

There are a few points to consider as the car world enters a state of flux. Vehicle buyers can no longer assume the market will remain the same as it has for the past 70 years. Buyers will have to look down the road and, although cloudy, some trends are becoming apparent.

- What kind of resale value will a new ICE vehicle have in 3, 6 and 10 years?
- What will fuel costs be in 5 years?
- What will licensing costs be?
  - Road taxes are collected from gas sales but that won't be practical with EVs. Taxes will likely have to be collected through licensing.
- What will maintenance costs be?
- What do you really think you will get for a used ICE car when you go into a dealership and all of the new cars are EVs?

  - This trend will be in full swing in most markets by mid-decade.
  - When motor cars began to replace the horse, it is likely that people tried to trade their horses as part of the purchase of the car. The same awkward negotiation of offering outmoded equipment as partial payment for modern equipment might begin to hold true for ICE-EV swaps with one large difference; horses were never made illegal. (not completely true)
  - But there is a strong possibility that ICE cars will suffer, if not complete banishment, severe penalties in terms of operating costs, licensing fees and fuel restrictions. There will also likely be restrictions of movement beginning with densely populated urban cores where ICE vehicles might be banned completely.

EV Trucks

The expression "If you got it, a truck brought it." nicely illustrates the ubiquitous presence of trucks, which are mostly diesel. Trucks consume approximately 20% of the fossil fuels we use for transportation. Unfortunately, diesel smoke produces ultra-fine particles which can enter the bloodstream causing pulmonary problems[35] in addition to all of the other issues associated with internal combustion engine exhaust.

Diesels are the preferred type of motor for trucks and heavy equipment given their durability, efficiency and low end torque. Although it was previously believed that electric motors and batteries could never replace diesels in industrial strength applications, that attitude is changing.

Electric automobiles are more reliable, more efficient and have much better low end torque than those with gasoline engines. Why couldn't they compete with diesels? Well, as it turns out, they can.

The diesel manufacturer, Detroit Diesel produces a model called the ePowertrain[36] rated for 180 hp and a staggering, 11,500 foot pounds of torque. Three battery sizes are available ranging from 210–475 kWh. Cummins, famous for its diesels, is offering a range of electric drive solutions for school and transit busses as well as transport tractors.

Readers on this planet may also have heard of the Tesla Semi.

Short range delivery vans are an ideal application of electric drive systems. They are excellent for stop and go milk runs. Van models from Nissan, VW and Maxus with 150–250 km ranges show a 75% reduction in fuels costs plus lower downtime and maintenance. The drivers also like their convenience, silence, smoothness and responsive driving characteristics.

Electric Flight

EV airplanes are much less of an ideal application because of the added weight of the batteries. This makes long distance electric flight and cargo carrying something which might appear only after many decades of technological progress. Large aircraft were made to take off with heavy tanks of jet fuel but they weren't designed to land regularly with them. Since batteries do not lose weight when they are depleted and currently have approximately a 15:1

---

[35] Diesel exhaust particles in blood trigger systemic and pulmonary morphological alterations, Abderrahim Nemmar, Ibrahim Inuwa

[36] https://www.trucks.com/2021/02/08/daimler-trucks-epowertrain-electric-freightliners

weight disadvantage over av. gas, commercial intercontinental electric flight will be a long time coming.

Heavy Equipment

Construction equipment suffer no such weight limitations and below are the remarks of a city manager on the use of EV excavators.

"We are collecting a lot of experience from this first pilot zero-emission construction site (which happens to be right outside our office building). There are three diggers operating there; two battery electric and one cable electric. The largest battery electric one (25 ton excavator) has an engine power of 122 kW (165 hp) and a battery of 300 kWh (400 hp). It can operate about 5 h on a full charge and when they charge during lunch break they can operate it for a full work day. The feedback from the construction company is that this works very well and they have managed to keep normal progress in their project. I also talked to one of the machine operators who praised the working conditions; less vibration, less noise and less local pollutants."

Is it so surprising that small and quiet electric motors can drive big machines? When you think about it, an EV being driven along the highway at 120 km per hour for hours on end might be using a continuous 30 kWs, or about 40 horsepower. Under hard acceleration, that could increase to 200 hp.

An excavator capable of lifting 20 tons might have a 170 hp diesel engine. A medium sized farm tractor might have an 80 hp engine. Electric cars now have motors many times as powerful.

Farm tractors[37] present a slightly different set of problems. Although they may be idle for long periods, when they are needed during narrow windows of opportunity for plowing, seeding and harvesting, they, like the farmers who drive them, might be called up on to operate almost 24 h a day. Power won't be the problem but range and recharging might be.

---

[37] Farmtrac 25G Electric Tractor 21,000 UK Pounds vs 13 k for diesel
http://www.escortstractors.com/

EV Motorcycles

The idyllic motorcycle experience is cruising through a pastoral scene of rolling hills, farms with green fields dotted with farm animals, the wind in your hair and the sun on your face with your motorcycle loudly spewing 1950s levels of pollutants behind it and causing animals to run away, babies to wake up and cry, and people to cease conversing.

An even better experience would be to do the same in silence to the point where the rider can hear their surroundings and smell the fresh air while everyone else simply continues to enjoy their day uninterrupted. The electric motorcycle has arrived.

Electric motorcycles present all of the same advantages and trade-offs as electric cars but they benefit even more from the characteristic silence and smoothness of electric motors. And compact electric motors offer new opportunities for designers to change the form factor of the two wheeled machines. The muffler and exhaust disappear as does any radiator. Weight again will be more due to the batteries but the penalty will not be as much as with automobiles (Fig. 9.29).

Electric motorcycles offer power, responsiveness, smoothness, simplicity and serenity. Plus some style if the designers see fit get creative. What they don't offer yet is very long range and fast charging. So touring on an electric motorcycle remains a carefully planned affair.

Zero is the best known electric motorcycle but there are many now coming on board as the motorcycle industry seeks to revive itself. Damon is a newcomer while the Sondors Metacycle is aimed at a niche offering a maximum

Fig. 9.29 Sondors electric motorcycle

speed of 130 km/h, a 130 km range (but not at 130 kph). It weighs 100 kilos has a 12 hp motor and 4 kWh battery and sells for $5000 US. Is this a perfect recreational and commuting bike? Time will tell. If it isn't, new designs will be taking its place to shake up the industry.

Electric Assist Bicycles

Electric assist bicycles and small scooters are revolutionizing urban transit and sport and recreational cycling. Some of the advantages are:

- Extend comfortable range knowing you will be able to get back easily.
- Don't arrive at work in a lather
- Keep up with the pack – an equalizer
- Carry shopping that was a load beyond what an unassisted rider could handle comfortably
- Make the bicycle the transport of choice.
- Fun – enjoy more and longer rides. Exercise is less strenuous but more extended and regular.

Electric assist bikes typically sense the force the rider is applying to the pedals and augment it with a certain amount of electric thrust. The level of assist can be varied considerably to take into account terrain and the effort the rider is willing to make. As well, most offer a throttle which allows the rider to rest completely.

Ranges can be 40–80 km depending on how much the rider pedals and on wind and hills along the way. There is literally an ebike for every possible application and kits which will add electric assist to a regular pedal bike in a matter of an hour.

Electric Boats

Halfway between a boat and a bicycle, the Manta 5 is a human/electric powered hydrofoil. Again, the compactness of electric motors opens up new function and design possibilities.

Nowhere do electric drive systems come into their own than on the water. They combine the speed and manoeuverability of power boats with the silence and smoothness of sailboats. Adoption of electric outboards is rapidly growing as people want more than ever to enjoy a quiet day on the water (Fig. 9.30).

Fig. 9.30 Manta hybrid electric hydrofoil

Electric outboards allow people to get away from noise, gasoline smells and exhaust fumes. Boating allows people a great level of physical freedom and it also gives them a chance to get closer to nature. This is a great deal easier using an electric drive system than a conventional outboard.

As with EVs, there is an abundance of information on range, speed and running time from both amateur and professional sources. Marine equipment dealers like Annapolis Hybrid Marine have built up a great deal of information from various sources as they've seen the electric outboard market (EOM) mature and slowly come into its own.

The electronic wave now appears to be feeding off itself as, according to Sally Reuther of Annapolis, "new electric car technology such as Tesla has made a lot of people sit up and take notice". Perhaps in the early days light bulbs paved the way for electric stoves which evolved rapidly. Now, as Reuther notes, one brand of electric outboards – ePropulsion – has actually offered regen capability to its product line.

The cost of an electric outboard is still higher than an internal combustion outboard but the prices aren't as far apart as one might assume.

As an example, a Honda 9.9 outboard might cost $2600 US whereas its electric counterpart, an ePropulsion Navy 6 with 6 kw of power – rated as a 9.9 hp – would cost $2200 and a 4 kWh battery $2000. The 2 kWh battery is $1200. On 4 kWh, the Navy 6 could power a small aluminum boat at 6 km per hour for close to 4 h (Fig. 9.31).

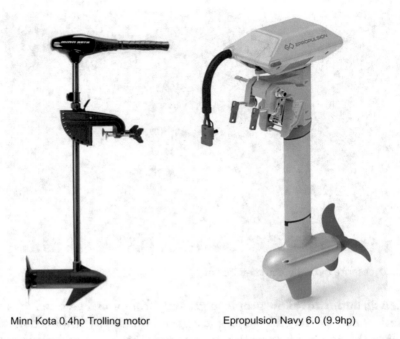

Minn Kota 0.4hp Trolling motor   Epropulsion Navy 6.0 (9.9hp)

Fig. 9.31   Evolution of electric outboards

As with their land-going counterparts, low rpm torque allows electric outboards to plane faster and, of course, throttle control is always going to be finer with an electric system.

Of course, this is perfect for fishing enthusiasts who crave speed and quiet. This video[38] gives a competitive recreational fisherman's point of view of the advantages of electric outboards. For more information on this rapidly developing area, contact an experienced dealer[39] just as you would for EVs. People who were on the bandwagon before it started moving can offer a great deal of valuable perspective. Perhaps Annapolis sums it up with their catchphrase *"Quiet power for your boat. Clean power for the environment"*. That is definitely where the world is headed (Table 9.3).

For reference the ancient and ubiquitous Minn Kota Endura C2 Trolling Motor, produces 30-lb of thrust and tops out at 0.4 HP while the newest and biggest electric outboards have grown to 80 horse power. There will certainly be more to come as EVs have demonstrated the lack of limits for electric drive systems.

[38] www.youtube.com/watch?v=OhL55rFNHsg

[39] www.annapolishybridmarine.com

Table 9.3 Runtime for Navy electric outboard

Power (Watt)	Speed (mph/kph)	Running time (hh:mm)	Range (mile/km)
500	4/6.5	18:00	72/116
1000	5/8	9:00	45/72
2000	6.7/10.8	4:30	30.2/48.6
3000	8/13	3:00	24/39
4000	11.5/18.5	2:15	25.9/41.7
5000	13.5/21.8	1:50	24.7/39.8
6000	15/24.3	1:30	22.5/36.5

Somewhat outside of the normal EV ethos and not known for their environmental or community friendliness, electric Seadoos can now be had offering 60 km of range and top speeds approaching 100 km per hour. Two models with 90 kW and 135 kW motors are rated at 120 hp and 180 hp respectively.

EV Tools and Landscaping

Anyone who has maintained a large property using an assortment of gasoline powered tools and tractors will attest to the fact that after a few years, they are spending a lot of time maintaining their tools. Gas lines have to be replaced, sparkplugs changed and both two stroke and 4 stroke gas containers must be refilled as well as stabilized as winter sets in. Corded outside electric power tools have been available for decades but powerful battery based units, while comparatively new, are now arriving in droves.

Electric snowblowers are expensive. But, as one reviewer[40] put it:

- Starts every time, no gas, no oil, high torque, no sparkplug or gas lines to replace, lighter and more maneuverable, no need to keep gas cans in the garage or shed, quieter than a gas snow blower (but far from silent), throws further, don't have to breathe exhaust fumes
- Downside battery life – pays to have spare batteries.
- An additional benefit is electrics can be hung on a wall.

From battery electric snow blowers to chain saws and lawn tractors, new technology and lower prices have launched these new designs past the bare threshold of acceptability to highly desirable.

---

[40] https://www.youtube.com/watch?v=Q2GhoDQa6tA

These products change the experience from one of working in a fog of high emission exhaust and noise to one with clean air and much less noise. The instant torque of the electrics also may require slight changes in work habits: be sure to have a firm grip on your chainsaw when you initially tip into the throttle.

These new tools typically have batteries which can be swapped out and charged inside. The batteries are expensive so it makes sense to commit to one tool line with compatible batteries across its range of products. Initial purchase costs are still higher than 2 and 4 stroke engine products but long term energy and maintenance costs will be lower.

As with EVs, battery electric tools have, in a single decade, gone from fairly marginal outdoor tools to mainstream products which out-perform their gasoline counterparts in most respects. If you haven't found an electric version of whatever tool you need, you probably haven't been looking hard enough.

Lawn Mowers and Tractors

So far, battery electric mowers look like their ICE forbearers but they have a feature that no gasoline mower can offer: the batteries can be used in your impact drill, chain saw, sander, router and weed-wacker.

Cordless power mower reviews are now plentiful such as this one which includes a "range – area cut" by "A Concord Carpenter".[41]

Electric commercial lawn mowers are replacing high pollution small ICE units, particularly 2 strokes. How many cars does it take to produce the pollution from one lawn mower? One site[42] claims running a leaf blower for 1 h produces as much pollution (not including carbon) as driving a car 100 km.

To convert fully to electric, contractors of all descriptions have to take extra steps to keep their machines working on the jobsite (Fig. 9.32).[43]

Snowmobiles

There are examples of conventional snowmobiles being converted to electric but here with the Envo, once again, we see the design flexibility that battery electric affords designers and manufacturers (Fig. 9.33).

---

[41] www.youtube.com/watch?v=lmKEZFtAAnI

[42] https://www.burlingtonelectric.com/lawnmowers

[43] https://www.blueroof.org/

Fig. 9.32 Makita electric lawnmower

Fig. 9.33 Electric snowmobiles - the greatest challenge for batteries?

Public Attitudes and Barriers

Many lists of "EV Myths" have been published over the years to counter public misconceptions. As well, many studies of public attitudes have been done in order to understand how to convince people to embrace this new technology. These may have been useful in the past but public sentiments and product development are moving so quickly that it isn't necessary to convince doubters to see the light. Any reluctant individuals can be left on their own as the world moves by them and they can grab on when their awareness comes up to speed.

The systemic corruption that actively tried to derail the transition to electric vehicles is largely now, itself, off the tracks. But there are large obstacles to face in the transition. Whole sectors of the economy have to be re-tasked and re-tooled.

Productive individuals in automotive repair, oil production and a wide range of industries will have to be transitioned into growing sectors. This will involve training, support and possibly re-location for hundreds of thousands of workers over years.

Fossil fuel production, storage and distribution facilities will have to be largely dismantled and cleaned up. In many, non-energy producing countries, these costs will not be large but in counties like Canada and the US, costs will run into the hundreds of billions of dollars.

Time to Catch the Green Wave

In 2010, the battery cost of the Nissan Leaf with 24 kWh was $1200 per kWh while in 2021, that cost had fallen to $130 for the Tesla 3. Future price drops will be more gradual certainly but they are still occurring. In early 2022, electric vehicles are the best long term choice for most people and their advantage over internal combustion vehicles will only increase in the coming years.

Technology improvements such as solid state batteries will, according to Toyota and VW, "increase the range of its vehicles by 30 percent and reduce their recharge time to 12 minutes by 2030." This approaches the fill-up time of gasoline vehicles. Such a technological leap forward would go a long way towards making gas cars a distant memory. Still, this is press release information and not a fully functional pre-production prototype.

But basic physics and conservation culture always has to be at the core of any green initiative as exemplified by BMW which declared that its new vehicles will focus on aerodynamics to maximize range. They will also assure sustainability in the full lifecycle of manufacturing, recycling and using renewable energy throughout the whole process.

The adoption of electrification is not something that has to be timed like catching a train or a ship. This process is now so well entrenched and complete that committing to it is more like getting on a conveyor belt. You can get on at any time you wish but the earlier you do so the further down the road you will be towards assuring your own energy resilience and reducing your carbon footprint.

The electric world offers more dispersed, localized generation and storage. Given the efficiency of electricity, there will be much lower gross energy consumption and vastly reduced carbon emissions. Converting to electricity and a lower footprint is not the horror filled story of deprivation it was once held out to be. There will be changes but, for citizens, they will mostly be positive. For society and for the planet, they are essential (Fig. 9.34).

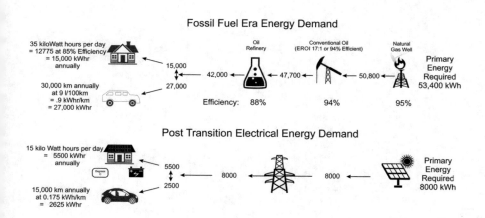

Fig. 9.34 Twenty-first century grid - getting the job done with far less energy consumption

10

Communities Lead the Way

What Kind of Communities Do People Want?

During the Covid-19 pandemic, those urban residents with the means to do so voted with their feet and sought health and safety in the cleaner more natural environments of rural or small town settings. There they could walk uncrowded trails and sidewalks and breath much cleaner air.

Aside from the higher infection potential in a dense urban environment, the flight to health and safety had another aspect: a return to nature. The pandemic has reminded many of us what has been missing in our lives.

Many urban dwellers have sought out nature in the past several decades but this became a national obsession when the pandemic struck. Crowded lobbies, elevators, offices and common spaces pushed people out of the cities and into safer and more natural environments which are healthier and simply feel better. The languages may be different but the meanings and the descriptions of the experience are the same.

- In Japan, "Shinrin Yoku" means "forest bathing" or "taking in the forest atmosphere".
- *The Norwegian word for* "nature cleansing" and "nature purification" is "naturrensing".
- German combines Wald ("wood") and Einsamkeit ("loneliness") together. Waldeinsamkeit literally translates to "solitude in the forest," but that literal translation maybe loses the word's poetry. "I'm most happy with life when I'm bathing in the world around me. Taking the time to slow things down and look…feel, smell and remember."

© The Author(s), under exclusive license to Springer Nature Switzerland AG 2022
J. E. Meyer, *The Post-Pandemic World*, https://doi.org/10.1007/978-3-030-91782-1_10

- Hygge - Hygge is a Danish and Norwegian word for a mood of coziness and comfortable conviviality with feelings of wellness and contentment. Possibly it means the same as feeling environmentally "well chuffed".

Was 2020 the year that the well-to-do abandoned the rat race and went "Hygge". Will Hygge be dropped in favour of gourmet coffee shops, commuting and crowded elevators and offices once the pandemic winds down? For some the answer is yes, but do we really believe this will be the last pandemic in our lifetimes? Possibly Covid-19 has changed our outlook on life.

Orillia is a small city of 31,000 in Central Ontario, Canada. Prior to the pandemic, the city council commissioned a study on citizen priorities and asked "How would Orillians spend their own money on improving life in Orillia?"

The public response is below.

Ranking:

1	Quality of life	26.5
2	Healthy environment	24.3
3	Sustainable growth	19.4
4	Professional progressive city	16.2
5	Vibrant waterfront	15.5
6	Heritage core	14.7

The total of quality of life and healthy living related categories was 82.5 (26.5 + 24.3 + 16.2 + 15.5).

The total involving any kind of simple growth was 19.4 but even sustainable growth implies quality, not quantity. Vibrant Waterfront (15.5) does imply more infrastructure.

These preferences are probably typical for virtually every town and city in the world. If people wanted larger cities, they could easily move to them but in many developed countries, there is now flow of citizens out of the large cities. During the pandemic, this became a flood. And in most cities, there is a drive to add more nature back into the urban environment, save farmland and reduce congestion.

Leading Communities Big and Small

Sustainable city initiatives go hand-in-hand with the new priorities of health, safety and nature with many advanced projects underway in many countries. Few countries are as urbanized as Japan where 92% of the population lives in cities.[1] Japanese planners are using the innovations of the Fourth Industrial Revolution to create new smart cities. But there is also a "Sustainable cities for a sustainable Japan" drive which sees cities operating under different human well-being focused mandates.

Fujisawa residents are the first consideration with city designers laying out a 100-year vision which took energy, security, mobility, wellness, community and even emergencies into consideration. According to Euronews,[2] "the city has environmental and energy objectives linked to CO_2 reduction, water savings, renewable energy use and, most importantly, a recovery plan in case of a natural disaster. They have made sure the city can be autonomous in electricity and food for three days."

Stockholm has been named the world's smartest city[3] as it aims for a carbon positive footprint by 2040. "Through innovation, openness and connectivity, we are making Stockholm more economically, ecologically and socially sustainable, from streetlight operation to mobility." According to Maria Holm project manager.

The organization encouraging smarter cities is named, unsurprisingly, "Smart Cities World". Their mission is to be the world-leading platform for sharing ideas to solve urban challenges that enable us to live in more resilient, sustainable, safe, and prosperous environments. The SmartCitiesWorld website[4] provides comprehensive and current global resources for urban leaders and their partners to find technology, concepts and working models for reference and comparison.

"Through our platform, we encourage the smart city ecosystem to share their experiences and learn from each other."

The Sustainable City[5] is a proposed project in the United Arab Emirates which "takes a three-tiered approach involving social sustainability, environmental and economic sustainability (Fig. 10.1).

The city of "Shining" Shenzhen[6] currently has 16,000 electric buses & 22,000 electric taxis which makes it world-leading. Not only is the hardware impressive, but the manner in which the whole system has been thought out

---

[1] https://www.un.org/development/desa/en/news/population/2018-revision-of-world-urbanization--prospects.html

[2] The city has environmental and energy objectives "linked to CO_2 reduction, water savings, renewable energy use and, most importantly, a recovery plan in case of a natural disaster." They have made sure the city is autonomous in electricity and food for 3 days.

[3] https://smartcitysweden.com/stockholm-named-worlds-smartest-city/

[4] https://www.smartcitiesworld.net/

[5] https://www.thesustainablecity.ae/

[6] https://www.youtube.com/watch?v=0P7fTPLSMeI

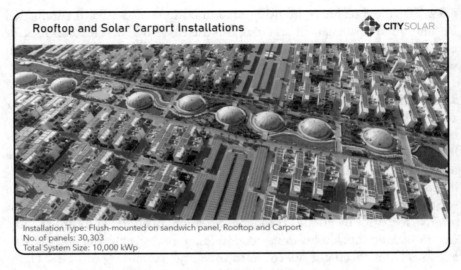

Fig. 10.1   Solar city with very high rooftop utilization

and supported by integrated infrastructure at all levels is equally key to their success. The air is cleaner, people are happier and the services are significantly improved while working conditions have also been enhanced.

For comparison, currently, the largest battery-powered electric bus fleet in North America is Canadian.[7] Toronto's transit system is now running 59 electric buses and hopes to have its bus fleet completely electrified by 2050.

Lahti, Finland, a very northern city of 120,000, has been voted the European Green Capital[8] for 2021 by instituting social innovations, not just technical ones.

What programs has it initiated?

1. 'Environmental nannies' educate children

These are elderly volunteers who go into class and help pass on their wisdom. They may also take them bird watching, build a hide-out in the forest, learn the flowers of a nearby meadow or even take them to a frozen lake for ice fishing.

Environment granny' Helena Juutilainen does not give orders. Instead, she interacts with the kids and shares the knowledge 74 years of life experience has given her.

---

[7] https://www.cbc.ca/news/technology/electric-buses-transit-1.5823166
[8] https://www.euronews.com/2021/02/19/european-green-capital-2021-six-climate-friendly-ideas-to-learn-from-lahti-in-finland

"We need to prepare citizens of the future to live a sustainable life," she said.

This program could be entitled "Generational Transfer" which takes culture building back from the digital world and hands it to those with life experience.

2. High-profile institutions show the way

The Lahti Pelicans want to be the world's first carbon-neutral ice hockey team. Their stadium is cooled with renewable energy and the club has scrapped air travel for getting to away fixtures. That means when they play in Oulu, about 500 km further north, it requires almost a day of travel on a sustainably fuelled bus.

"The big idea here is to set an example," Pyykkö said. "We are in a position where we can have a great impact on ice hockey fans, other players and society in general."

The Lahti Symphony Orchestra has worked towards becoming carbon neutral and has partnered with researchers from the Lappeenranta-Lahti University of Technology to make sure that emission cuts are counted following the latest scientific guidelines.

3. Reward citizens for climate-friendly everyday actions

Reduce your carbon footprint and receive a reward: that is the logic behind the launch of a personal carbon trading app.

Based on answers to a survey about your current life situation — asking how many children you have, how long your commute to work is and similarly themed questions — the app calculates your personal CO_2 emission budget.

"If you manage to stay under budget, you are rewarded virtual euros that you can use for tickets to the swimming hall or the bus or to buy a puncture repair kit for your bicycle," project manager Anna Huttunen from the city of Lahti explained.

The world's first personal carbon trading app has been tested in LahtiCredit: Lassi Häkkinen/City of Lahti.

It is designed to educate and inform and take climate action on a personal level. Think of it as a carbon selfie.

4. Make money from sustainable innovations – Lifecycle connections

Sustainable thinking is not only good for the climate and environment. It can also make businesses bloom, as the members of the Päijät-Häme Grain Cluster in the Lahti area have seen. The cluster is an extensive network of companies working with grain in all links of the production chain, from farm to retail resulting in one project which saw the innovation of a bread bag made of 25% oat husks.

Claim back polluted areas. Remediation is widely needed for garbage dumps, industrial sites, public dumping grounds and over 100 years of careless disposal habits. Now, the glittering water invites you to come for a swim and to eat high-quality fish from the lake. The restoration of Lake Vesijärvi was one of the first of its sort, and it has since become a model for other similar projects in Finland and abroad.

5. Invest in the best.

Transitioning from coal to biowaste. A new plant burns certified biomass, consisting mainly of wood chips from forest industry residues and by-products. Together, the new plant and the Kymijärvi II plant, which pioneered the generation of energy from waste by gasification in 2012, supply the majority of Lahti's 120,000 residents with electricity and district heating. With a 70% decrease in carbon emissions, "Sure, we could have found a cheaper solution by burning fossil fuels like natural gas, but this is an environmental investment and well worth it."

Note the grassroots and broad based nature of the Lhati initiatives!

Corporate driven examples exist such as Toyota's[9] high-tech "Woven City". This fosters a human-centric approach to community and will be powered by zero-emission technology. The guiding themes are 'human-centred,' 'a living laboratory' and 'ever-evolving' where inventors and innovators solve problems for communities in a timely fashion.

Provincial and state governments are important and New York State[10] takes the lead according to researcher Art Hunter who lists the key elements of their efforts:

1. The five focus points for both retrofit and new build are:

    (a) Energy efficiency
    (b) Renewables integrated onto the grid
    (c) Grid harmony – collaboration with building owners and grid managers

[9] https://driving.ca/toyota/auto-news/news/toyota-breaks-ground-on-its-high-tech-woven-city
[10] https://youtu.be/WRu3s0Cy6iY

(d) Building codes are vital but integrated with GHG awareness, appliance codes, material codes, manpower training, building inspectors, developers, and more. There are many standards to blend holistically.

(e) There is embedded carbon in everything mankind is doing. How to reduce this and maintain a happiness or well-being index is a BIG issue.

2. Technical

(a) New York is developing codes and technology for district heat pumps. This means 10+ homes or 2+ big buildings as a minimum forms a community.

(b) NY has 4.5 million buildings and needs to upgrade 400/day to meet 2030 targets.

(c) Using new and upgraded roads for imbedding ground source heat exchangers.

(d) Getting rid of gas stoves and appliances is vital for indoor healthy air quality.

3. Funding and jobs

(a) Going Green is not just for the wealthy - Environmental justice to every citizen. Nobody left behind.

(b) Issues like training for new jobs, existing pensions, health, technology and many cross cultural, cross industry, cross departmental, cross education course content at all levels – in other words a fast reacting and flexible social safety net.

(c) Critical strategy of pathways and tracking progress against the plan. Monitoring is essential.

(d) The rate structures need to adjust for time to incentify low cost through lifestyle adjustments etc. Awareness is vital through training, use of media, etc. And full cost life-cycle rather than only capital cost accounting must be used.

4. Governance coordination is vital for leadership since regulatory policies across departments require holistic approaches. Timing matters.

New Jersey wants to avoid a repeat of the disaster which saw a 150 km/h derecho trigger a raw sewage spill[11] into their Woodlynne Borough. The plan is to develop a microgrid which can maintain power for nearby hospitals, schools, police stations and other critical infrastructure in times of emergency.

---

[11] https://www.nj.com/news/2021/03/camden-county-needs-a-backup-plan-for-future-storms-clean-energy-can-be-the-solution-developer-says.html

There will be more discussion of the Drake Landing development in Okotoks, Alberta in the Geothermal Section but it consists of 52 homes linked to a geothermal storage "heat battery" which derives its heat from banks of solar hot water panels. The development has reduced external energy consumption by over 90%.

The above working examples illustrate the many different ways leading communities are making strides in both reducing their footprints and improving the lives of their citizens.

Is Where You Live, Livable?

A book that deals with this exact question is "America's Most Sustainable Cities and Regions: Surviving the 21st Century Megatrends" by John Day and Charles Hall. It "details how humans can take control of our collective destinies in spite of potholes in the road such as the Great Recession of 2007–2009 and the pandemic."

"Our current attitudes were formed during a unique 100-year period of human history in which a large but finite supply of fossil fuels was tapped to feed our economic and innovation engine. Today, at the peak of the Oil Age, the horizon looks different. Cities such as Los Angeles, Phoenix and Las Vegas are situated where water and other vital ecological services are scarce."

"Finite resources will mean profound changes for the energy-intensive lifestyles of the US and Canada but not all regions are equally vulnerable to these 21st-century megatrends. What are the critical factors that will determine the sustainability of your municipality and region?"

The book addresses critical points such as:

- How resource availability and ecological services shaped the modern landscape.
- Why emerging megatrends will make cities and regions more or less livable in the new century.
- It ranks cities on a "sustainability" map of the United States.
- It finds that urban metabolism puts large cities at particular risk.
- It concludes that sustainability factors will favor economic solutions at a local, rather than global, level.

Although the book is written for the US, the discussion it puts forward is applicable for any region on the planet.

Take a hard look at where you live and determine whether it is a stable and broad based community. If your community is not sustainable, neither are you. Can your community be upgraded to become flood and fire and drought resilient? This has to be designed into a community and this article[12] "Architects are designing for the new climate reality" discusses some of the ramifications of climate change for infrastructure.

Your Town: Latitude Matters (Again)

Latitude still plays a huge part in the costs of assuring a stable energy supply. Using rough estimates, below we examine the size and cost of solar PV systems for two very different cities.

A single family residence in Igloolik in northern Canada and well above the Arctic Circle might optimistically have a 40 kWh annual average daily heating budget and need 120 days of storage during the winter since the solar array would produce almost no energy. In the summer, it would have only 120 days of full production to cover current daily requirements and charge up the batteries for drawdown in the winter. During the remaining 125 days in the fall and spring periods, the energy produced would roughly match energy demand.

The variables of maximum production and demand change by latitude so for the southern reference city of Wilmington, North Carolina there are 320 production days and only 4 days of storage needed to assure reliable delivery of its much more modest 12 kWh daily budget.

The panel arrays have to be large enough to allow for daily consumption plus filling the storage batteries in the period of the year when there is strong sun. At the extremes again, Igloolik requires a large 63.4 kW array (254 × 250 W panels) at a cost of $127,000. In Wilmington, a single family dwelling could be provided with the heat it requires and the household power needed with a 3.2 kW array at a cost of $6300.

Storage, if utilizing Tesla PowerWalls at a cost of $850 per kWh, would cost $4.1 million for that single family home in Igloolik and $41,000 for the family in Wilmington.

Also in Wilmington, upgrading to an 8 kWh array would provide enough energy to allow an efficient EV to travel 40,000 km annually (Table 10.1).

Most of the population of Europe and Canada live north of Wilmington and south of Igloolik which puts their necessary investment higher than

[12] https://www.cnn.com/style/article/flooding-architecture-projects-spc-intl-c2e/index.html

Table 10.1 The local climate really does matter for the ease of decarbonization

	Annual kWh budget	Storage days required	Annual storage kWh required	Size of array required	Storage	Array	Total investment
Igloolik	14,600	120	4800	63.4	$4,080,000	$126,736	$4,206,736
Wilmington	4380	4	48	3.2	$40,800	$6337	$47,137

Wilmington's but still much lower than that required for an Arctic Circle location. A very large proportion of the world's population live south of Wilmington and their typical Total Investment might be even lower than Wilmington's.

This math is valid as a Survival Tip for Preppers but they have to keep in mind that survival in a long term crisis can only come as a part of a healthy and resilient community. Attempting to survive on one's own in a bunker comes with an expiry date. Only a socially cohesive, broad-based community can assure the long term stability necessary to provide adequate food and energy supplies along with health and education facilities.

Healthy Communities

To the World Health Organization (WHO), "A healthy city[13] is one that is continually creating and improving physical and social environments and expanding community resources. These enable people to mutually support each other in performing all the functions of life and developing to their maximum potential."

A Healthy City aims to:

- to create a health-supportive environment,
- to achieve a good quality of life,
- to provide basic sanitation and hygiene needs,
- To supply access to health care.

What is the ultimate goal? See "Hygge" or perhaps become familiar with the Bhutanese Index of Happiness. The WHO stresses continuous improvement by forging the necessary connections in political, economic, and social arenas rather than setting static minimum standards.

Can your local political leaders commit to this program and adopt these Five Healthy Cities governance principles?

- Integrate health as a core consideration in all policies
- Address all social, economic and environmental determinants of health.

  - implement urban development planning and policies which reduce poverty and inequity

---

[13] https://www.who.int/healthpromotion/healthy-cities/en/

- address individual rights
- build social capital and social inclusion
- promote sustainable urban resource use

- Promote strong community engagement.
- Reorient health and social services towards equity: ensure fair access to public services.
- Assess and monitor wellbeing and increase transparency accountability.

As we have come to understand much more clearly because of the pandemic, public space, clean air, access to nature and uncrowded public transport can lay the groundwork for public health and security in good times and in bad.

Grids

Grids are the regional and national power distribution systems which currently supply you and your community. In the future, they will also take your excess energy production and re-distribute it. Electrical and possibly heat energy grids are going to become much more complex and more flexible as we transition out of the simplicity and built-in storage capacity of fossil fuels. Power generation with a large renewable component will be more variable and needs to be buffered to provide consistent energy and voltage for the end user. And ever fluctuating consumer demand will have to be modified as much as possible to accommodate and exploit the variability of solar and wind resources (Fig. 10.2).

Clearly having fleets of V2G capable EVs which charge at night and can discharge as needed, along with smart washing machines and dryers which operate when rates are low, will flatten the fluctuating demand curve. Geothermal storage, pumped storage will do this on a much larger scale. Necessarily, electricity distribution needs to get smarter as do consumption patterns.[14]

In principle, smart appliances could form a spot-trading network of their own within the home, queuing up for priority access to electricity based on the prevailing price, demand, and the owner's preferences. An average household should be able to cut its electricity costs and carbon footprint by an appreciable amount simply by delegating to appliances the agency to manage the timing of their own operations.

---

[14] https://spectrum.ieee.org/energy/the-smarter-grid/electricity-distribution-needs-to-get-smarter-on-a-finer-scale

Variable Inputs

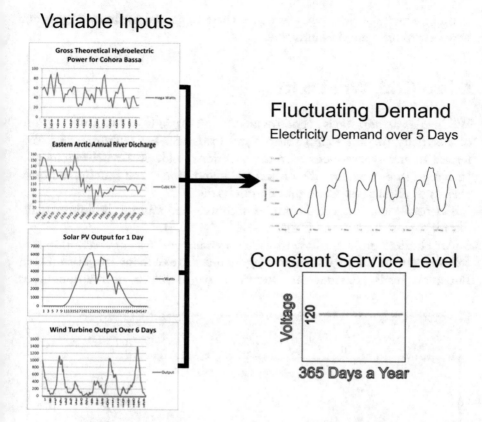

Fig. 10.2 Grid designers have their work cut out for them with highly variable inputs and customer demand for high service levels

Will the assignment of decision making to appliances end in the fridge telling the family when they can and cannot buy ice cream? This very large question is one which humanity must grapple with as we build increasingly complex systems which will actually be monitoring our activities and making decisions we may not be conscious of.

Art Hunter's living lab geothermal/solar PV/heat pump/V2G EV system consists of three main computers and a number of smaller Raspberries requiring a huge amount of expertise to build and operate. Will owners need to understand the operation of their system in broad strokes for it to perform well or will all decision making be done by the system?

Perhaps individual owners will chose the level of involvement they are comfortable with. Some will chose to operate their home like a sailboat where everything has to be manually adjusted, Some will prefer a more autonomous setup in which the rider tells the horse where to go and the horse figures out how best to navigate the landscape.

Micro-Grid, What Is It?

Wikipedia's description is as good as any. "A microgrid is a decentralized group of electricity (or hot water) sources and loads that normally operates connected to and synchronous with the traditional wide area synchronous grid (macrogrid), but can also disconnect to "island mode" and function autonomously as physical or economic conditions dictate."

Essentially any connected complex of sources and multiple buildings qualifies for the catch-all term of "micro-grid" (Fig. 10.3).[15]

Art Hunter's project, whose mission statement and system description is as follows, sets the bar for a small microgrid and also as a living lab. The mission statement reads "Retrofit an Ottawa home to live a near autonomous

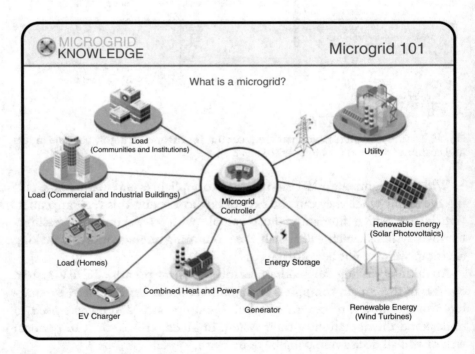

Fig. 10.3   An idealized microgrid

---

[15] https://microgridknowledge.com/microgrid-defined/

lifestyle." For the rest of us, this can be taken to mean creating a resilient energy system for safe and consistent home operation.

"The Manotick microgrid hardware includes 16 kW solar power generation (PV), 40 KWh Lithium Ion home storage batteries, a small V2G capable Electric Vehicle and a low grade heat supply with a Ground Source Heat Pump. After 5 years of integration with Internet of Things hardware and software, the verification program has demonstrated a 300 consecutive day period without buying electrical power or any fossil fuels" (Fig. 10.4).

Dr. Hunter's system buries waste heat underground in a Svec loop geothermal system which has a depth of approximately 3 m. Average ground temperature in Manotick, 20 or so kilometers south of Ottawa, is around 12 °C and by the end of the winter, the temperature of ground around the system has dropped to near zero Celsius. Over the spring and the summer, Mother Earth naturally brings the temperature back up aided by any excess heat from the house. Hunter does not currently have any solar hot water arrays to charge the system to more than ambient ground temperature.

Resiliency is the key to social stability. The mini-crisis of the Colonial Pipelines ransomware attack that reduced gasoline supplies on the US east coast for about 10 days, demonstrated the vulnerability of large systems, to partial or complete failure. One small wrench in the cogs of energy supply can cause significant personal and economic disruption for scores of millions of

Fig. 10.4 A microgrid in the flesh

people. Home energy systems and micro-grids increase resiliency and reduce the threat of widespread damage no matter what the cause.

Microgrids allow individuals and small groups or communities to exercise their initiative and build stability, resilience and low, long term cost into their lives without depending on government or utility involvement. Microgrids also allow essentially an open architecture of technologies and scale for projects which are attractive to a wide range of investors.

And who wants to invest in microgrids aside from homeowners? Apparently everyone according to a Green Tech Media report.[16]

"Increasing system standardization and the declining costs of energy resources have reduced development costs and boosted the growth of small microgrids." This makes due-diligence costs lower. The bottom line is solar PV and geothermal systems are no longer bleeding edge technologies, they are well proven with highly predictable results.

District Heating

The invisible battery. Under front yards, back yards, parks, roadways and parking lots.

District heating is a group of buildings all connected to one or more sources of heat energy or storage. It is basically a micro to medium sized grid that generates, stores and transfers heat rather than electricity.

The following is condensed from the book: "The Renewable Energy Transition, realities for Canada and the World".

There are many local opportunities to implement cogeneration strategies employing district networks for heat distribution. Gas plants, incinerators, and some types of industrial processes produce substantial quantities of heat which can be captured and distributed by a hot water network among dozens or hundreds of nearby houses and commercial buildings. In Sweden, garbage that cannot be recycled is saved for the winter months and burned, with the end energy output breakdown being 10% electricity and 90% heat. In regions with suitable geology, there are opportunities for geothermal storage with deep drilled piping storing very large amounts of energy hundreds of feet underground for use months later by nearby buildings. Heat sources can range from solar hot water systems all year round or the heat "effluent" from air conditioners in the summer, whereby the geothermal

---

[16] https://www.greentechmedia.com/articles/read/everyone-wants-to-own-a-microgrid

system would both make cooling in the summer more efficient and store the heat for later use.

Widely used in Europe, district heating systems are estimated to save up to 40% of energy costs for those connected to it. Of course, the energetic cost of the source dictates the ultimate cost, but if the system uses waste heat, or a strong geothermal source, then costs can be kept quite low.

There are over 5000 district heating systems in the United States supplying 8% of the heat used in commercial buildings.

A variant of this is the district cooling system in use in the central core of Toronto which uses the coldest water, around 4 C, from Lake Ontario to cool scores of large buildings near the waterfront before the water is then fed into the municipal water system. This system saves 90% of the air conditioning costs of conventional systems although one would have to ask about the effect on the health of Lake Ontario if such systems were to remove significant quantities of its coldest water (Acciona 2019).

Drake Landing

By far the most advanced and proven development in North America is the Okotoks, Alberta community of Drake Landing utilizes solar PV and hot water panels to harvest energy and borehole thermal storage. All 52 houses are serviced by insulated piping. An array of solar hot water panels generate 1.5 MW of thermal power during a typical summer day and supply heat to the district heating system. Surplus heat is stored in a collection of 144 holes extending down 37 m below the surface. The temperature of the earth will reach 60 °C by the end of each summer. With over a decade of data, this project can serve to inform the design of similar projects in any local.

This process has proven to be, on average, 45% efficient over the 10 years of its life. These essentially conventional 140 m^2 (1400 ft^2) homes still have an average daily energy budget of 35 kWh but, although very high by global standards, it is significantly lower than that of the average Canadian house.

In Canada, space and domestic water heating account for more than 80% of greenhouse gas emissions in the residential sector; hence, the emissions from these houses will have been cut by 72% (Fig. 10.5).

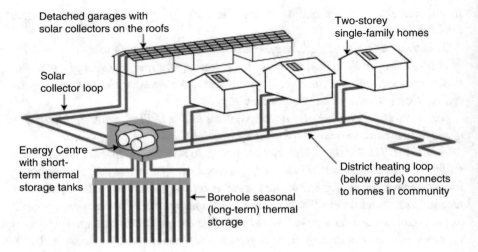

Fig. 10.5 The Drake Landing development is a starting template for energy positive housing in northern climates. (Drake Landing Solar Community: 10 Years of Operation, Lucio Mesquita et al.)

Cogeneration

Cogeneration describes an energy generator which creates more than one stream of energy. Typically this might be a natural gas plant which was designed to produce electricity. Burning fossil fuels produces heat and previously the excess heat was literally sent up the stack to be exhausted into the atmosphere. But when this large amount of waste heat is harvested and either distributed around the neighbourhood or stored in the ground, the process qualifies as "cogeneration".

District heating from industrial sources include:

- Incinerators
- Foundries
- Large office complexes
- Server farms
- Hockey and curling arenas - ice is created by taking heat away.
- Body heat from commuters or fans at large sports arenas

Drake Landing is a simple district heating system but such systems can involve many sources which can be added in over time.

Geothermal Heating, Cooling and Storage

Geothermal Storage

The counter-intuitive concept of pushing heat into the ground and pulling it out months later has been proven to work in many regions of the world, foremost in Sweden. Soil formations have to be suitable which means there cannot be significant water circulation around the geo-heatmass carrying heat away.

Subterranean energy storage will be most important for northern climates, but regions which experience extreme heat could also benefit from the use of geothermal cold storage. Low-grade energy can be used to heat buildings as the application does not require very concentrated energy or even extremely hot water. The Okotoks hot water system operates at between 30 °C and 60 °C. This minimizes losses and allows enough heat to be delivered to the houses through its hot water grid.

Geothermal is a simple system, whose most complex mechanical part is a pump for water (glycol). The total cost for the solar hot water and storage first-of-a-kind Drake Landing was $4.3 million or $82,000 per house.

The cost of a geothermal heat pump system starts at $25,000. In the Drake Landing project, the thermal mass of the storage is effectively 20,000–30,000 m^3. When the temperature of this mass is raised to the 70 °C peak, it represents storage of over 1.5 million kWh of energy or 30,000 kWh of storage per house. If that amount of storage was provided by electrical batteries, say our Tesla PowerWall reference, 2300 of them would be required for a cost of over $2 million per house. The cost of chemical batteries is dropping but will probably never approach the cost level of geothermal storage now available for the $80,000 and certain to become cheaper.

In effect, as a heat battery, geothermal storage never goes completely flat. One can overdraw one's thermal account by drawing the temperature down below the natural 10 °C ground temperature level, but, over time, Mother Nature will bring the account back up to zero. Electric batteries do not work like that and neither does your friendly neighbourhood bank.

The effective cost of the geothermal storage capacity is $5.00 per kWh, bringing a geothermal equivalent to a 13 kWh, $10,000 PowerWall in at a cost of $65 or 0.6% of the cost of electrical storage. With geostorage recyclability is probably just maintenance. Geostorage is best applied to district heating-sized projects, but clearly it can also work well for individual homes and buildings since geothermal systems are in wide use now.

By virtue of their permafrost underpinnings, the most northern communities will not be able to use geothermal storage as it would require drilling perhaps 100 m down to hit a stratum that is warmer than zero Celsius. So those communities will have to remain fossil fuel dependent while using renewables for summer and immediate use. The very powerful storage characteristics of fossil fuels simply can't be replaced by renewables in all cases. "Populate the North" campaigns need to have an energy reality check.

As lifecycle analysis begins to reveal the full long-term costs of our activities, the low investment cost and negligible recycling and maintenance costs of borehole geothermal storage will become increasingly attractive.

At the other end of the scale are the cooling problems which beset much of the lower latitudes.

Cooling will become more critical as the climate warms and extreme heat events become a mortal threat to hundreds of millions of people. A wet bulb temperature is a measure of temperature and humidity combined. Our internal body temperature is 37 °C and skin temperature is kept below 35 °C allowing body heat to be lost through the skin.[17] The wet bulb temperature of 35 °C is the temperature and humidity condition at which we are unable to shed heat. Death occurs once our temperature approaches 42 °C. Permanent damage can occur before that.

Geothermal cold storage is nothing new. During the summer we've stored large blocks of ice in cold cellars and barns in beds of sawdust for centuries. These blocks were then dug out and transported to facilities or homes where they "powered" cold boxes – early refrigerators - and provided primitive but effective and very green air conditioning. Now we can send cold fluid down pipes to store the cold (reduce the heat) in a mass of earth for months and then bring it back up when the hot weather calls for it (Fig. 10.6).

Curtailment

Curtailment is the dumping of power that cannot be handled by the grid when, for instance, strong winds during a warm night create high wind turbine output that exceeds demand. The more storage available throughout the system, the lower the curtailment losses will be.

Curtailment can be expected to drop over time as more storage and more progressive time-of-use regimes are developed. Cheap electricity would be

---

[17] (Sherwood & Huber, 2010)

Fig. 10.6   Heat and cold storage goes back many centuries

welcomed at almost any time by thermal storage systems which could always use another degree or two of ground temperature.

If those ground batteries match the Drake Landing system efficiency level of 45%, then prices down into the range of several cents a kilo Watt hour would be extremely attractive. And geothermal storage can soak up a great deal of energy very quickly. So can a large fleet of parked EVs. Simplicity is another attractive characteristic of geothermal storage. Given favourable ground conditions, geothermal storage systems are simple to build, simple to maintain and simple to recycle, if that term even applies here.

But clearly, the challenge to respond to changing demand and generation levels is a problem that will be front and center for utilities going forward. They will be looking for fast response energy sources, and the two major ones will be hydro and natural gas. Both of these can ramp up in as little as 5 min. For smaller scale fluctuations, electric batteries and hydrogen could also fill in very quickly.

Buildings

Lee Kuan Yew was Singapore's visionary leader who oversaw Singapore's transformation into a developed country with a high-income, high equality economy within a few short decades. His core policies formed a system of meritocratic and anti-corruption government. Lee eschewed narrow interest policies in favour of long-term social and economic planning.

The Lee Kuan Yew School of Public Policy stresses transparency, education and holistic leadership.

Lee was the driver of Singapore's "Garden City" initiative.[18] CNBC reported that Lee "introduced that idea in 1967 when he was prime minister. Its elements included roadside greenery, featured most prominently along the East Coast Parkway (ECP) highway, which connects the city-state's international airport to the city center, with colorful bursts of tropical flowering shrubs and imperiously-tall trees" (Figs. 10.7 and 10.8)

In Singapore, utilizing all surfaces on roofs, and facades for solar PV electricity production could supply 20–25% of the city's electricity demand.[19] In terms of geothermal cooling, air conditioning systems would be much more efficient (higher COP) if they were ground-sourced using 25 °C earth than if they were air-sourced using 30–40 °C air.

A large geothermal development in Toronto called "The Well"[20] will use a multi -million litre column of water ("well") below it as a thermal battery. It will also draw off the cold of the city's drinking water being taken from the bottom of Lake Ontario to assist cooling.

Creative designs optimized for the local conditions range from wild to woolly... Clearly all regions should be determining the optimum design of

Fig. 10.7   Singapore makes buildings part of the living environment

---

[18] https://www.cnbc.com/2016/03/27/lee-kuan-yew-was-actually-singapores-chief-gardener.html

[19] https://www.youtube.com/watch?v=PM101DvvG4Q

[20] https://www.cbc.ca/news/business/climate-heat-cooling-1.5437701

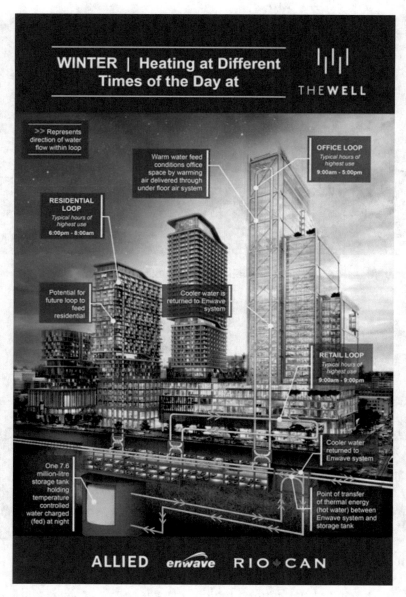

Fig. 10.8 Mainstream developers come on board with geothermal

buildings best suited to their own unique climate and energy challenges. One size or design does not fit all (Fig. 10.9).[21]

---

[21] https://www.euronews.com/green/2021/05/06/the-10-most-tagged-eco-friendly-buildings-in-the-world

Fig. 10.9 Brazil architectural innovations make public spaces come alive

Fig. 10.10 Milan Twin Towers

Rio's Museum of the Future Outdoor manages outdoor shade and indoor light. With some of these designs, the question of maintenance arises and clearly instances of freezing rain are rare in Rio. These high functioning and feel-good buildings provide a core of similar benefits (Fig. 10.10):

- Fresh air
- Ventilation
- Open space
- Shade / Natural light
- Low noise
- Exposure to nature

The Milan Paired Towers feature built in fresh air but is maintenance an issue?

Community Leadership

Green Education and local coordination events bring enthusiasts, activists and policy makers into contact with the mainstream public (Fig. 10.11).

From electric cars to DIY solar, these events can energize local greening efforts. Some politicians are afraid to attend these events because it would be an implicit admission that climate change and finite fossil fuel reserves are real threats. But the more people talk about these issues and support educational events and projects, the less politicians will be intimidated by business-as-usual brow-beating which is now fading into a fringe.

Politician's lament: scientists and citizens say this but my pocketbook says that.

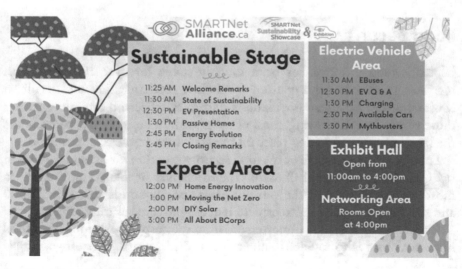

Fig. 10.11   Local initiatives drive progress

Renewable Energy Infrastructure Planning

Endless growth and sustainability are mutually exclusive and nowhere is this made more clear than in analysis of the scale of the infrastructure needed to replace fossil fuels. Instead of getting energy from invisible subterranean oil-fields hundreds or thousands of kilometers away, renewable energy infrastructure is all above ground and highly visible.

It has to compete with farmland and existing infrastructure. Certainly roofs can be put to great double duty with energy harvesting but, at the end of the day, renewables will simply require a great deal of local space. Besides the biophysical impossibility of sustaining current levels of consumption, further growth directly reduces the quality of life citizens seek as well as the future resilience and stability of their communities (Table 10.2).

Although local politicians crave growth and some businesses are dependent on it, the quality of life for actual residents declines with the rate of growth of a city or town (Fig. 10.12).

Growth in the size of the commercial market isn't the answer for the future because it hasn't been the answer for over 60 years as the graph above implies. In fact, quality of life and life satisfaction have been declining in most western countries for decades while debt has skyrocketed despite very strong economic growth.

Check your city's rank[22] in terms of traffic congestion, one of the strongest negative barometers of community health and individual well-being.

Food Security

Globally, foodland area has been roughly stable for over 100 years with increases in output coming from fossil fuel, mechanization and fertilizer. Most countries have seen populations start to stabilize and even decline. For those which are still growing, most of the growth in housing to accommodate the larger numbers is taking place on the very finite resource of farmland.

Following is an example of the food security question for one small town in the Canadian province of Ontario. The mechanics of this example apply to most towns and cities of the world. Orillia's population is currently 31,000 and the local council plans to expand it at a rapid rate basically forever.

Ontario is a net energy importer and is on the cusp of becoming a net food importer. That is to say, it is vulnerable to disruption in these two critical

---

[22] https://www.tomtom.com/en_gb/traffic-index/ranking/

Table 10.2 Conflict between growth and the quality of life and social stability

	Impact of growth	Impact of quality and stability
Budget	Higher demand for infrastructure, higher taxes, deficits	Lower demand for additional infrastructure, upgrading of current infrastructure, balanced budgets
Jobs	More jobs means nothing if they don't pay a living and tax positive wage, lower wage gig jobs have higher service demand	Stable number of increasingly high quality jobs expands tax base with lower services demand
Natural space	More people reduces natural space per person	Natural space is preserved
Healthy living	Higher density is less healthy as Covid-19 has demonstrated	Healthy environment, healthy living and healthy jobs increase quality of life and reduce service demand
Equality levels	Increased inequality stresses social cohesion and reduces quality of life for all income levels	An egalitarian society offers higher quality of life for all income levels and has lower support services demand.
Affordable housing	Growth means inflated housing costs, higher inequality, debt and lower quality housing	Young families can afford to live near their parents and raise children in safe neighbourhoods
Food security	More people means more food demand and reduced agricultural land due to urban sprawl	Foodland preserved and food demand stabilized allowing focus on healthy food
Renewable energy transition	Renewable energy sources are diffuse and require large collection areas. The larger the population, the more area must be "industrialized" with energy infrastructure.	A stable population with greater investments in efficiency, conservation and stability will minimize the size of the renewable energy infrastructure.
GDP	Larger GDP size serves those who live off asset inflation and the flows of money.	GDP per capita indicates the economic health citizens.

aspects of long term social viability and resilience. Orillia is located at the northern fringe of the main body of Ontario's agricultural land. Consequently, the expansion of Orillia will be done mostly on agricultural land as it is for most cities in the world. The carbon emissions from paving one hectare of land are in the range of 10–20 tonnes.

Using wheat as a proxy for the entire diet consisting of 2800 calories per day, the number of hectares of farmland is laid out for various levels of population in the table below (Table 10.3).

The above table assumes the average person consumes 2800 calories a day, 365 days a year and that there are 3500 calories in every kilogram of wheat

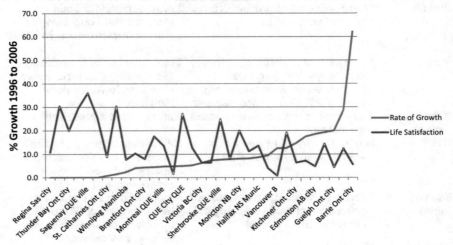

Fig. 10.12 Congestion, unending construction and unaffordable housing apparently haven't found favour with citizens

Table 10.3 Foodland area required for larger populations

Population	Total food energy demand at daily calories of 2800	Annual calories (billion)	Annual kg of wheat at 3600 calories per kg	Hectares of wheat at 3300 kg per ha
15,000	42,000,000	15.3	4,258,333	1290
30,000	84,000,000	30.7	8,516,667	2581
41,000	114,800,000	41.9	11,639,444	3527
80,000	224,000,000	81.8	22,711,111	6882
120,000	336,000,000	122.6	34,066,667	10,323

and the average yield is 3300 kg per hectare. Of course, a 100% wheat diet would be very unhealthy and a healthy diet would require many times the cropland area that wheat would. Wheat was chosen to demonstrate the minimum area of cropland that would be required to support a given population.

Food security depends on the ratio of foodland per capita and once this falls below a certain level it assures that the region in question must get food from outside its catchment area. Being dependent on imported food presents similar but more severe risks than being dependent on foreign sources of medical supplies. At some point it is going to emerge as a real problem and paving over farmland is a certain means of guaranteeing that the problem will be worse than it had to be.

Table 10.4 Renewable energy infrastructure required to support larger populations

45° latitude Canadian city population	Daily energy demand (kWh)	Hectares of farm at 0.078 ha/ capita	Solar farm (Km²)	No. of 2 MW Turbines at rate of one turbine per 100 people	Hectares of wind farm at 0.548 ha/ capita	Wind farm (Km²)
15,000	2,250,000	1170	11.7	150	8220	82.2
30,000	4,500,000	2340	23.4	300	16,440	164.4
41,000	6,150,000	3198	31.9	410	22,468	224.7
80,000	12,000,000	6240	62.4	800	43,840	438.4
120,000	18,000,000	9360	93.6	1200	65,760	657.6

The solar PV and wind infrastructure necessary for Orillia, to meet the challenge of climate change and the transition to renewable energy, will take up many thousands of hectares of farmland. This infrastructure will necessarily cover a very much greater land area than our current energy generation sources and take some foodland out of production just as effectively as urban sprawl (Table 10.4).

How many hectares of farmland are there currently in Orillia's catchment area? How many will there be once the solar and wind generation infrastructure has been installed? As well, if the population is increased, how many hectares will be lost to urban sprawl?[23]

The above simple illustrations must be fleshed out by comprehensive mathematical models but they give an idea of the issues and the scale of the effort needed to achieve sustainability. Clearly, the ambition of a small minority to continue with population growth is rooted more in greed than in responsible analysis. Are politicians in your area able to respond to these fundamental issues? Are they including food and energy resiliency in their planning?

Does Orillia have a strategic analysis of the infrastructure necessary to transition to renewable energy and feed itself at even its current level of population? Orillia, like all jurisdictions, needs a mathematical model to provide ongoing guidance and act as a transparent base for discussion. Models have been completed for a number of communities and they can be customized for any municipality.

---

[23] The analysis of Ontario's ability to feed itself is fleshed out over a shorter term in this study commissioned by the Ontario Farmland Trust: "Farmland Requirements for Ontario's Growing Population to 2036 - Charlotte McCallum, PhD". A more extensive study looks at the overall farmland issues for Ontario AGRICULTURE AND AGRI-FOOD ECONOMIC PROFILE FOR THE GOLDEN HORSESHOE by Margaret Walton

Municipal modelling software by (CIS) CityInSight[24] has been implemented for many Canadian cities and there are models developed for use in a number of countries.

An additional resource is the UN site – Make Cities Resilient 2030[25]– which offers a very useful checklist to enable movement forward with strategies to achieve resiliency and sustainability. For instance, the City of Ottawa produces only 5% of the energy it consumes which shows how far some jurisdictions have to go to achieve resilience and why individuals need to assure their own home is well prepared for exceptional events.

Social cohesion is much more important for the resiliency of renewable systems than for our current systems. A nuclear generating plant with 4 mega-Watts of capacity might take up a square kilometer and be fairly simple to secure. A disrupted society might find it difficult to prevent looting and vandalism on the many 50 hectare sized solar installations scattered around the countryside it would take to replace a nuclear power plant.

Recycling and Lifecycling

Lifecycle analysis – yes, it is complicated. Lifecycle analysis is the measurement of all inputs and outputs throughout the chain of supply involved in the production and use and finally the recycling of any given product. It involves everything from how much energy it took to pull the minerals out of the ground to how much energy it will take to recycle the end product into useful raw material (Fig. 10.13).

Without recycling, lifecycling becomes irrelevant and your community eventually becomes a giant waste dump. How well your community recycles is a very clear indicator of how sustainable it will be in the future.

Community Projects

Local gardens and forests are often the lifeblood of community involvement. Below is a list of considerations for pushing these projects forward.

Botanic Gardens Conservation International 7 Golden Rules[26]:

---

[24] http://cityinsight.ssg.coop/
[25] https://mcr2030.undrr.org/
[26] https://www.bbc.com/news/science-environment-55795816

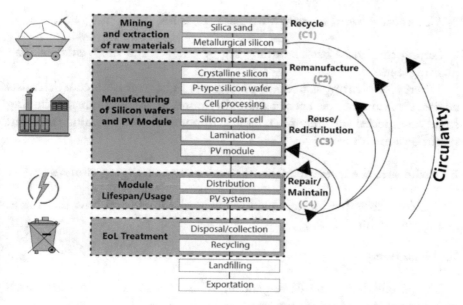

Fig. 10.13 Lifecycle material flows. C.C. Farrell. ("Technical challenges and opportunities in realizing a circular economy for waste photovoltaic modules" C.C.Farrell)

1. **Protect existing forests first**

Keeping forests in their original state is always preferable; undamaged old forests soak up carbon better and are more resilient to fire, storm and droughts. "

2. **Put local people at the heart of tree-planting projects**

Studies show that getting local communities on board is key to the success of tree-planting projects. It is often local people who have most to gain from looking after the forest in the future.

3. **Maximize biodiversity recovery to meet multiple goals**

Reforestation should be about several goals, including guarding against climate change, improving conservation and providing economic and cultural benefits.

4. **Select the right area for reforestation**

Plant trees in areas that were historically forested but have become degraded, rather than using other natural habitats such as grasslands or wetlands.

5. Use natural forest regrowth wherever possible

Letting trees grow back naturally can be cheaper and more efficient than planting trees.

Where tree planting is needed, picking the right trees is crucial. Scientists advise a mixture of tree species naturally found in the local area, including some rare species and trees of economic importance, but avoiding trees that might become invasive.

6. Make sure the trees are resilient to adapt to a changing climate

Use tree seeds that are suitable for the local climate and how that might change in the future.

7. Make it pay

The sustainability of tree re-planting rests on a source of income for all stakeholders, including the poorest.

One of the great tragedies of the twentieth century was the loss of the American chestnut to an Asian blight. Through careful and relentless cross-breeding, the American Chestnut Foundation is hoping to bring back this magnificent species to play an important part in the regeneration of forests in the American North-East. As the climate warms, its range can now also extend north into Canada.

This quote from the ACF website[27] lends scale to the potential for carbon fixing and economic benefit of a chestnut revival.

"More than a century ago, nearly four billion American chestnut trees were growing in the eastern U.S. They were among the largest, tallest, and fastest-growing trees. The wood was rot-resistant, straight-grained, and suitable for furniture, fencing, and building. The nuts fed billions of wildlife, people and their livestock. It was almost a perfect tree, that is, until a blight fungus killed it more than a century ago. The chestnut blight has been called the greatest ecological disaster to strike the world's forests in all of history.

The American chestnut tree survived all adversaries for 40 million years, then disappeared within 40."

Native tree species are falling prey to foreign bugs and changing climate and either supporting them or replacing them with useful and resilient species is critical to the health of forests around the world.

---

[27] https://acf.org/the-american-chestnut/history-american-chestnut/

Restoring forests is a great project for any area. If you are selecting a new species for your area take carbon fixing and nut or fruit output into account. Trees can offer economic benefits throughout their lives, not just when they are cut down. Annual harvests keep the community involved and aware of the importance of their forest resources.

Promoting farmers markets, walking trail construction, car and bicycle sharing programs and widespread public charging will make the community more connected and healthier.

Other projects can range from creating a living energy lab at a local school and cleaning up past environmental damage through landfill and waterway remediation. Re-establish wetlands and natural waterfronts to give nature a toe-hold to become self-sustaining.

Soil Re-building

What we have done to the environment extends to the ground under our feet (Lowdermilk: Conquest of the Land Through 7000 Years[28]). Bringing back the health of the soil is a prime choice for personal and community initiatives and should be a national priority. The strength of the planets carbon sinks is declining as forests burn off and the capacity of the oceans to absorb more is now in question. Building soil sinks carbon.

The photo below shows two handfuls of soil, the left being the original sand/silt slightly acidic Canadian shield soil and in the right, the same soil with 2 years of simple composting of fresh wood chips with plants growing in that soil and hosting the biology that was doing the magic (Fig. 10.14).

Restoring soil is best done with a plan. According to Sundaura Alford-Purvis of the Society for Organic Urban Land Care: "Five things that I think that everyone in a decision making position about land has a responsibility to do before altering it would include:"

- Learn the history of the land and the life that the land hosts/hosted
- Learn the ecosystem that it exists within
- Learn how it contributes to and relies on the surrounding ecosystem
- Learn about their own history of relationship with land and ecosystems
- Learn how humans have historically supported the capacity of that land, or similar ecosystems, to support life

---

[28] https://www.nrcs.usda.gov/wps/portal/nrcs/detail/national/about/history/?cid=stelprdb1043985

Fig. 10.14 Two handfuls of soil, one degraded and the other remediated

Many native species are being eliminated by invasive bugs and climate change throwing into question which species re-planting should include. Alford-Purvis also points out that when selecting tree species for re-forestation, it is best to choose those which are currently thriving in climate regimes perhaps a thousand kilometers to the south because that is the climate regime which will be arriving in your area over the next few decades. This concept is fleshed out on the Shelter Wood site.[29]

Models and Sources

The Rocky Mountain Institute's handbook[30] details cobeneficial actions regional governments can take to limit the effects of climate change and improve society.

---

[29] https://www.shelterwoodforestfarm.com/
[30] https://rmi.org/insight/regions-take-action

The case studies from India, Brazil, Europe and the United States includes economic development, air quality, public health, equity and resilience in its model to illustrate how all regional governments can move forward.

- Commit to generating clean electricity. Renewables are cost-effective investments, and clean electricity is fundamental to a carbon-free society.
- Construct and upgrade carbon buildings to be all-electric and efficient, which will also create healthier, more comfortable places to live and work.
- Create healthier mobility options and electrify vehicles, which can reduce air pollution while giving people more choices for transportation.
- Convert industry to electric power while driving an innovative, clean energy economy.
- Make land use sustainable by stopping urban sprawl and enhancing the natural area that create beautiful places, economic opportunities, and essential carbon "sinks".

"The stories in the guide are exciting because they illustrate how governments are addressing very immediate problems while also leading the way on climate change solutions," said Jacob Corvidae, RMI principal and co-author of the guide. "There is so much that can be achieved for society as a whole when governments take a comprehensive approach to making their communities more climate-friendly."

Strategy – Building Blocks of a Community Renewable Strategy

The many examples above which have produced clear beneficial results are direct action initiatives. They had clear objectives and were defined by specific actions. This contrasts to programs of carbon offsets or cap and trade in which polluting parties continue to do what they are doing and buy green righteousness by buying the goodness that someone else is doing or is purportedly going to do.

Although they may build awareness, these programs are what amounts to licenses to pollute and don't involve direct reductions, or improvement. Even if they are well monitored and concrete results can be proven, they subsidize business-as-usual rather than directly drive change.

Community involvement spawns understanding and ownership and is critical for successful implementation of any program. The public has to know there is a problem and has to be made aware of the benefits of solving it.

Government openness to innovation means a willingness to take risks because not all projects work out as hoped. A great deal can be learned from successful projects but failures also bring their own useful lessons. The radical difference between the timeline most governments work with and the timelines of the biophysical world mean that something that might be deemed a political or economic failure turns into a social and biophysical success further down the trail.

The term "living lab" means that we are constantly learning and improving. The Global Commission on People-Centred Clean Energy Transitions[31] helps countries to advance their shifts to clean energy technologies by enabling citizens to benefit from the opportunities and navigate the disruptions.

Education is the core and key strategy of the transition. Living labs put the hands of children on energy systems to give them a basis for a lifelong association with energy, consumption and the world around them. Ultimately, an informed, involved and innovating populace is the future.

What Drives Individual Decisions?

Confidence enables decisions and budgets and robust models underwrite confidence. The City of Oslo was became the first to develop **a carbon budget.** It set a cap on emissions leading to carbon neutrality by 2030. Oslo managers find the carbon budget to be a very important management tool and it is consulted side by side with the financial budget.

The decline in consumption is usually driven by both price mechanisms and public policy priorities in the form of standards and incentives. If we have collectively decided that we want to continue to live on this planet then consumption of fossil fuels has to decline dramatically. Standards which curtail the use of certain materials or limit the emissions of certain pollutants need to be accompanied by a consistent price structure.

As people climb into affluence, their consumption levels increase. But at some point they tend to level off and once the lure of living large has given way to living well, consumption levels can begin to decline.

---

[31] https://www.iea.org/programmes/our-inclusive-energy-future?utm_campaign=IEA%20 newsletters&utm_source=SendGrid&utm_medium=Email

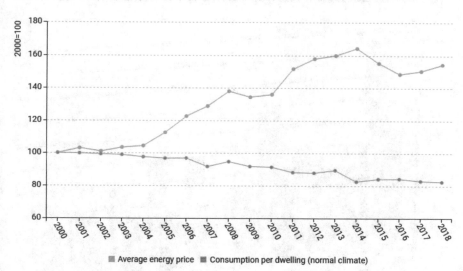

Fig. 10.15 When energy prices go up, people use less energy. When prices stabilize, so does consumption

Prices may be increased for the high carbon fuels to curtail their consumption and incentives can be used to support the substitution of other low or no carbon fuels. In mature countries, efforts to reduce energy consumption can start off strongly but once the low hanging fruit has been picked, progress slows as shown in Fig. 10.15.

Higher prices encourage strategies which replace expenditure on energy with investment in energy saving. But if fossil fuel prices level out and are possibly even surpassed by inflation, efficiency improvements will tend to slow (Fig. 10.16).

Progressively higher fossil fuel energy prices are necessary to underwrite consistent investment in clean energy. Large fluctuations in fossil fuel costs will only disrupt personal and corporate consumption as well as investment patterns. The ultimately result will be a much more fitful transition and much higher levels of stranded assets.

Enthusiasts, early adopters and green evangelists will form the bleeding edge of the renewable energy transition. Cost is secondary to these leaders. Most people, although willing to be converts, will not be active leaders and some, as we've seen with the clearest biophysical threat to come along in a

CO2 Embedded in Imports vs Domestic Emissions

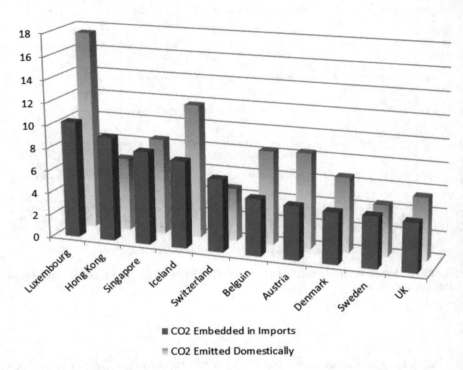

- CO2 Embedded in Imports
- CO2 Emitted Domestically

Fig. 10.16 Coal fueled China supplies the world with goods carrying a very large carbon price tag

century, will even remain staunchly stuck in the mud. In order for the sustainability transition to take place, price mechanisms will have to be aligned to push the process seamlessly along. Humanity does not have the decades to burn that a stop-and-start price rollercoaster would inflict on implementation times. Price signals have to be clear and consistent.

Budgets and Modelling, a Critical First Step

Sophisticated modelling software is necessary to assist policy makers and concerned citizens in dealing with the hoary problem of simultaneously changing to low carbon energy sources while reducing overall energy and material consumption over decades. This transition has to unfold without placing undue stress on groups less able to handle it.

EcoEquity[32] is a calculator designed to help policy makers avoid the social disruption that would go hand-in-hand with an inequitable carbon reduction program. It also lays out the total amount of carbon a jurisdiction might have to "spend" whether that budget is met in 3 or 100 years. This web based calculator can be down-scaled to determine total allowable carbon emissions for cities[33][34] and is designed to allow the user to specify their own preferred interpretation of national responsibility and capability for climate action.

The trajectory of the emissions reduction path matters, too. Climate change is a function of how much greenhouse gas we emit into the atmosphere *in total*, not just annually. If emissions rise too much before we hit carbon neutrality, we will not prevent catastrophic global warming. Some sources estimate that we will exceed our survivable carbon budget by 2029.

The idea of a budget is simple, and it is a familiar tool for planning at all levels of government.

According to Ralph Torrie of the Sustainability Solutions Group,[35] "A carbon budget provides an overarching framework for GHG emissions management, extending over multiple years and over all aspects of community social and economic activity. The carbon budget aligns with both science and with decision-making frameworks used by local governments for capital and operating budgets—frameworks in which investments, costs, and benefits are assessed over multiple years and often involve trade-offs between early action and deferred spending."

When combined with effective emissions monitoring, the carbon budget enables transparency and the feedback needed to make periodic adjustments to the budget.

An all-inclusive measure of CO_2 emissions has to account for net embedded CO_2 and energy. Total carbon footprint is a big reason to move manufacturing back to nations with lower CO_2 emissions and with shorter supply lines. What are the energy and CO_2 emissions embedded in the goods your nation's imports?

"Total allowable" budgets indicate that reducing emissions is more urgent than we anticipated because "the acceleration" is now well underway. The organization Carbon Neutral Cities[36] lists cities committed to carbon

---

[32] https://calculator.climateequityreference.org/

[33] SSG is working with Pathway to Paris to design and build a carbon budget tool based on these design principles that establishes carbon budgets for nearly 300 cities around the world. (1000 Cities Calculator)

[34] City of Ottawa's Energy Transition Strategy, City of Edmonton's Energy Transition Plan Update.

[35] http://www.ssg.coop/

[36] https://carbonneutralcities.org/cities/

neutrality and provides a great many examples of the myriad initiatives being undertaken to achieve that goal.

Carbon budgets[37] provide a scientific and systematic framework for mounting the emergency response that has become an imperative for communities everywhere. They also offer a useful means of comparing results. Since we live by energy and may die by carbon, budgets for each of these are far more important than the dollar based fiscal budgets all governments focus on. They need to work with the two most vital biophysical issues on humanity's existential plate in their own physical units.

Every jurisdiction needs to have energy and carbon budgets.

Lead by Example: Yours or Someone Else's

Your local leaders may or may not be very progressive in their outlook. No matter how keen or reluctant they may be to achieve sustainability for their community, they are unlikely to have a clear vision of what this entails if they aren't fairly far down the modelling and strategy trail.

If you can help them to visualize what a sustainable society looks like and the types of tools and metrics that are involved in managing in this new world, you will have made it much easier for them to implement positive steps towards achieving that goal. For individuals needing to be dragged into the future, being presented by clear goals and workable tools now in use in other regions leaves them much less able to deny, divert and deflect action on pro-resilience initiatives.

Getting ahead of the curve is expensive but still less expensive than being caught behind it when a threat becomes reality. The pandemic demonstrated that with brutal clarity and the biophysical threats, mostly of our own making, lined up behind Covid-19 have the potential to do a great deal more damage to the stability of our society than the virus has.

In the past, many municipal politicians have conflated growth with progress along the lines of "A Bigger Boston is a Better Boston". Towns and cities are human constructs to serve human interests and when a politician or journalist bestows human status on a city, they are really speaking about their own interests rather than those of the general public.

Any town or city pursuing a policy of population growth is unlikely to be serious about renewable energy, resilience or sustainability. They won't have the necessary goals, transparent policies or models in place to deliver a

---

[37] UK Cities Lead on Global Climate Action Goals We Need to Accelerate Canada's Green Recovery

determined and successful effort. More likely they will be dug in to obfuscate, deflect and delay to allow business-as-usual to keep the current cash cow healthy. The greater good is not on the table.

Growth advocates don't like clear goals, science or full sets of numbers. Citizens want a clean, healthy environment with easy access to nature, safe streets, short commutes and high quality, affordable housing. This is in direct conflict with the business-as-usual policies of many local politicians in western countries.

If your community isn't sustainable, neither are you. You don't want to be the only house with the lights on during the zombie apocalypse.

The voice of theory can make itself heard through protest and repetition but the voice of experience is the one people most respect and pay attention to. If you yourself have made changes to your home, vehicle and lifestyle to vastly reduce your environmental footprint and carbon emissions, people will want to know more. If you have worked with real numbers and can refer to them in conversations and presentations, your message will be much better received.

Leadership inspires followers whereas individuals exhorting people to do what they have yet to, will find audiences much tougher. A well-worn path to a known result is much more inviting than breaking trail over uncertain terrain towards an unknown destination.

11

Your Country; Sustainable, Resilient and Secure

National Government Responsibilities

Ultimately, national governments are responsible for the overall policies which will lead to both a successful transition to renewable energy and a restoration of environmental balance. They need to act in the following areas:

- Food Security
- Energy Security
- Establish clear national goals
- Establish metrics of social and biophysical health
- Assure commercial market decisions are energy and resource rational
- Monitor effectiveness of current policies
- Maintain coordination of activities of all levels of government

A responsible national government has to balance the influence of narrow interest groups with the common good and it must assure public perceptions match reality. In Canada, 70% of the public believe the country is a leader in the fight against climate change. But in fact, Canada has one of the worst records of ghg performance and of breaking their climate commitments in the world.

This disconnect makes it much more difficult for a responsible government to implement effective policies. The cold light of day will eventually wash over the "Sunny Ways" leadership approach and push it aside to allow real issues to be addressed in a substantive manner. But that means catching up on decades of shirked responsibility.

© The Author(s), under exclusive license to Springer Nature Switzerland AG 2022
J. E. Meyer, *The Post-Pandemic World*, https://doi.org/10.1007/978-3-030-91782-1_11

Generational Responsibility

What do the following have in common?

- The Japanese ethic of "son gets the better rice field"
- Iroquois planning horizon of seven generations
- The Good Ancestor[1]: Following the Intergenerational Golden Rule[2]

These societal ethos require continuous improvement and recognition of an adult's responsibility to the many generations which will follow after as well as the work and sacrifice of ones ancestors; a binding cultural compact.

Too little long-term thinking has been setting modern civilization up for failure. How often do most people think about the effect of their actions on the wellbeing of future generations? The last two generations of westerners have been taught that we should live and consume for today assuming tomorrow will be better. This is a huge break from thousands of years of human tradition.

According to author Roman Krznaric, "...humankind has always had the innate ability to plan for posterity and take action that will resonate for decades, centuries, even millennia to come. If we want to be good ancestors and be remembered well by the generations who follow us, now is the time to recover and enrich this imaginative skill" (Fig. 11.1).

Fig. 11.1 Growthbusters advocate being a good ancestor

---

[1] The Good Ancestor: A Radical Prescription for Long-Term Thinking – Roman Krznaric.
[2] Dave Gardner at www.growthbusters.org/good-ancestor/

Fig. 11.2   Cascading crises may leave future generations with no good options

If becoming good ancestors makes us bad consumers that is something our children can easily live with. Researcher Alan Garber in his paper Rights and Responsibilities vs Self-centredness[3] addresses the balance between immediate satisfaction and long term prosperity we must to arrive at if we are to lay a stable foundation for future generations (Fig. 11.2).

"Many are loath to bear the immediate costs of limiting carbon release into the atmosphere despite the growing awareness that eventually the failure to act will lead to environmental catastrophe. There are few rewards to legislators and government leaders who ask constituents to make sacrifices now to avert a future problem that seems as abstract and distant as it is potentially devastating."

"Tallying the cost of the pandemic can help in understanding how much society stands to gain by preparing for the next one."

The bottom line is; are we going to prepare as best we can or will we hand down a planet on fire to our children who will be unprepared, ill-equipped and overwhelmed?[4]

---

[3] Learning From Excess Pandemic Deaths Present Centeredness, Alan M. Garber.

[4] https://www.nbcnews.com/video/deadly-wildfires-sweep-through-algeria-s-kabyle-region-118357061880

Food Self Sufficiency

A food crisis is even more critical than an energy crisis unless the energy crisis strikes a northern town in the dead of winter. Many nations are now net food importers. Perhaps many of the foods they import are along the lines of interesting varieties which enhance the eating experience. But there are also a large number of cases where imports are used to assure a minimum daily caloric intake for their populations.

The very nature of agriculture has changed and so too has it relationship to consumers. Distant and energy intensive, with a heavy dependence on mechanization for production, transportation and storage, makes the food supply vulnerable to a variety of disruptions as this paper details "Declining Country-Level Food Self-Sufficiency Suggests Future Food Insecurities (Fig. 11.3).[5]

For an excellent overview of the foreign food dependence issue, the Overshoot Day Organization offers background and a listing for most countries.[6] As well, Jennifer Clapp's study "Food Self Sufficiency and International Trade: A False Dichotomy?" covers these questions in more detail.

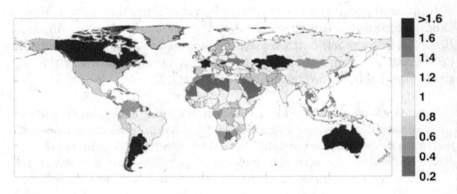

Fig. 11.3 How food independent is your country or region? (*Source*: Reproduced from Puma et al. 2015, based on FAO data. http://iopscience.iop.org/article/10.1088/1748-9326/10/2/024007/pdf)

---

https://media-cldnry.s-nbcnews.com/image/upload/t_focal-460x230,f_auto,q_auto:best/newscms/2021_32/3498232/210811-algeria-fire-mc-10252.JPG

[5] https://link.springer.com/article/10.1007/s41247-019-0060-0 John R. Schramski.

[6] http://www.overshootday.org/food-deficit/?utm_campaign=EOD%202020&utm_medium=email&_hsmi=93633324&_hsenc=p2ANqtz%2D%2DIhjWCR4fX3xD_pe65xENILgq1H-BAeNxV8svt5841sKj7yFCY5da_sB94-ZtIAbgoFqSh00sLmNOd6mC6mRa_cixrodlNunxNk2OJM1-m1vqRlP7E&utm_content=93633324&utm_source=hs_email

Food production balance is a very fluid dynamic varying greatly from year to year and crop to crop. But paving over farmland, no matter how great a country believes its surplus to be, is reckless.

The Whole Biophysical Enchilada

Global warming has justifiably seized the lion's share of public concern over a declining environment but what if climate change is only the tip of the melting iceberg?

Ecologist William Rees points out that the size of the human enterprise has grown to the point where our foot has substantially outgrown our ecological boot. We have displaced a large proportion of the natural systems we depend on and not only must we achieve carbon neutrality but our impact on the planet must be reduced in its entirety (Figs. 11.4 and 11.5).

It falls to national governments to develop complete representations and models of their countries' biophysical realities. If climate change is simply yet another symptom of ecological overshoot as Dr. Rees and numerous physicists and environmentalists contend, then our current emphasis on ghg emissions must be matched by efforts across a broad spectrum of issues rather than simply trying to clean up one end of the pond (Fig. 11.6).

Dr. Rees suggests that instead of using a maximum consumption model as the commercial economy requires, we construct rules for consumption and

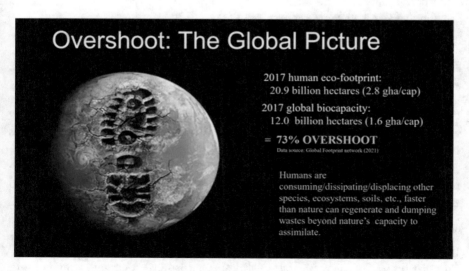

Fig. 11.4   Overshoot: the planet as our doormat

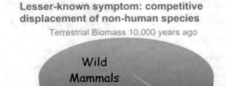

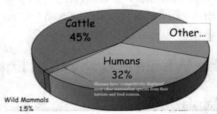

William Rees, Vaclav Smil

Fig. 11.5 Displacing the natural world with one based on fossil fuels

Climate change is merely a symptom of *ecological overshoot*

The human enterprise is using resources and generating wastes in excess of the regenerative and assimilative capacities of the ecosphere.

Some other symptoms of *overshoot* (excess economic scale)

- ❏ the oceans are acidifying
- ❏ fresh waters are toxifying
- ❏ the seas are over-fished
- ❏ soils are eroding
- ❏ deserts are expanding
- ❏ tropical forests are shrinking
- ❏ biodiversity is plummeting (etc., etc.)

Meanwhile, income gaps widen even as global wealth accumulates.

Fig. 11.6 Climate change is just one symptom of a complex problem. (William Rees, Canadian Club of Rome presentation)

our interaction with nature just as we have created a culture of rules for how we deal with each other to avoid conflict. Whole system environmental rules can establish an orderly and agreeable society which can perpetuate itself.

Certainly there are challenges ahead because the richness of fossil fuels cannot, in the foreseeable future, be re-created from renewable energy. And energy is the flip side of commodity availability.

Dr. Fatih Birol, IEA Executive Director[7] points out that: "Today, the data shows a looming mismatch between the world's strengthened climate

---

[7] https://www.iea.org/reports/the-role-of-critical-minerals-in-clean-energy-transitions?utm_campaign=IEA%20newsletters&utm_source=SendGrid&utm_medium=Email

ambitions and the availability of critical minerals that are essential to realizing those ambitions."

- Clean energy demand for critical minerals is set to soar
- Production of critical minerals essential for key clean energy technologies like electric vehicles and wind turbines need to pick up sharply over the coming decades to meet the world's climate goals, creating potential energy security hazards that governments must act now to address.
- Manufacturing the typical electric car uses six times the mineral inputs of a conventional automobile, and an offshore wind plant requires 13 times more than a similarly sized gas-fired plant.

If we have our environmental Pearl Harbour or Italian hospital overload moment, will we immediately come together and take the right corrective action? Or will we fail to identify the root cause of the cataclysm and just start fighting each other for scarce resources as we have done repeatedly in the past? Covid-19 was easy to identify and focus on as was Pearl Harbour. Scarcity and climate change have larger impacts but their fingerprints are smeared as is the accountability for fixing the problems. Is there a pathway to success?

Coherent Policy Connects the Dots of Cause and Effect

Identifying the root causes of problems is the core around which a strategy is built. The battle against Covid-19 was fought with changes to the way we live and through the development of vaccines. Vaccines give the body a familiarity with the virus and a toolkit to be able to deal with the real virus when it is encountered.

Human societies have been through many cycles of environmental decline and the resultant societal collapse. We tend to study the wars which typically spring from this process but instead we should be studying why the wars occurred. We should have learned, as our bodies have, how to deal with the familiar threats that are now, once again at our door.

Resource depletion, environmental decline and climate change are a set of threats that humans have contended with since we became a species and they have been documented in one form or another since we learned to record our thoughts.

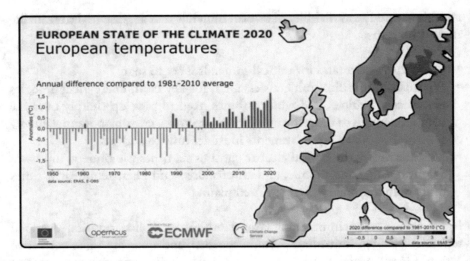

Fig. 11.7   Europe is warming quickly

The thousands of generations of humans who lived close to the land could clearly see biophysical threats coming and would take steps to both mitigate them and get out of their way. Our highly stratified society has seen the last 2 or 3 generations becoming largely isolated from the natural world and its processes. But we have developed an increasingly rich reserve of scientific expertise capable of identifying threats and connecting them to their causes. This knowledge, however, has yet to work its way into the media or political discussion enough to drive effective change (Fig. 11.7).

Instead of watching weather and migration patterns and the fertility of the soil, modern humans can quantify their observations and assemble a time series of data and present it in the form above to illustrate the point that our world is changing. The past 200 years been one of the most favourable periods for human life in the past several hundred thousand years and any change in climate must be considered a threat.

But now, at least we can document what has happened and predict that substantial changes are in store. This is analogous to the first 2 weeks of the pandemic where competent governments recognized there was a problem and began to ramp up testing and restrict travel.

Now that we are over the pandemic hump, most governments clearly recognize what the best approach would have been. Reports studying government response to the pandemic found that the absence of a coherent overview and in-depth strategy was fatal for government efforts seeking to protect the national interest. Rather than conflate commercial interests with the national interest, successful programs connected the dots and dealt with the problem on its own ground.

Underlining the common disconnects between expert scientific advice and political action is the US National Intelligence Agency's 2021 Annual Threat Assessment.

- "The effects of the COVID-19 pandemic will continue to strain governments and societies, fueling humanitarian and economic crises, political unrest, and geopolitical competition as countries, such as China and Russia, seek advantage through such avenues as "vaccine diplomacy."
- Ecological degradation and a changing climate will continue to fuel disease outbreaks, threaten food and water security, and exacerbate political instability and humanitarian crises.
- Warmer weather can generate direct, immediate impacts—for example, through more intense storms, flooding, and permafrost melting.
- We will see increasing potential for surges in migration by Central American populations, which are reeling from the economic fallout of the COVID-19 pandemic and extreme weather, including multiple hurricanes in 2020 and several years of recurring droughts and storms.
- The number of people experiencing high levels of acute food insecurity doubled from 135 million in 2019 to about 270 million last year, and is projected to rise to 330 million by yearend.
- The threat from climate change will intensify because global energy usage and related emissions continue to increase, putting the Paris Agreement goals at risk."

The NIA report clearly connects the cause and effect dots but will politicians recognize biophysical reality by enacting biophysical action programs? Can they recognize and deal with the simple conflict illustrated in the graph below through effective legislation? (Fig. 11.8).

Farmland is something all levels of government should be concerned with but ultimately, if the national government is promoting population growth then the loss of farmland and food security will be felt at some point in the future.

Many mature countries which have experienced cyclical food crises have implemented population and foodland strategies. Other, 'young" countries such as Canada, Australia and the USA still have a developer power class and colonial mentality focused on population growth and exploiting ever more natural resources.

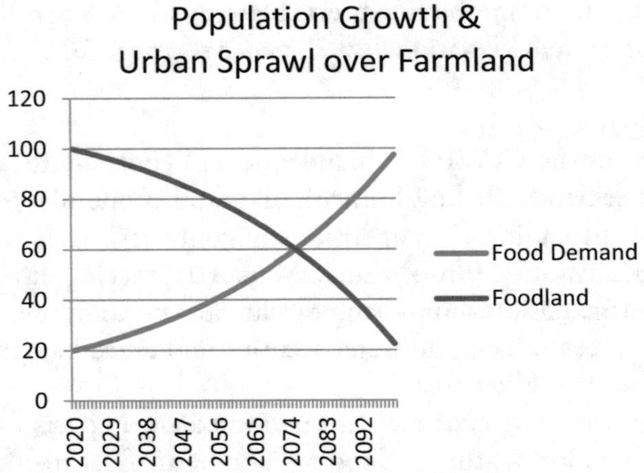

Population Growth & Urban Sprawl over Farmland

Fig. 11.8 The simple problem of sprawl vs farmland hasn't been grasped by some politicians

Connect-the-Dots Successes

• Every society which has lived within its means and not had to invade a neighbour or see its population migrate.
• Countries which experienced hardships and the limits of their own depleted resource bases and who have evolved to adopt conservation, efficiency and lower consumption levels.
• Countries which no longer have simple growth as the nation's focus but rather social stability and environmental balance.

Connect-the-Dot Societal Fails

• Every society which has invaded another or which has seen a significant portion of its population out-migrate or experience long term privation.
• Countries pursuing endless growth which will inevitably lead to the above outcomes.
• Countries using monetary metrics to determine social and environmental policy.

Failure ensues when people capitulate and throw their hands up resigning themselves to their fates due to circumstances that are overwhelming. They then look for miracles. There is a difference between depending on miracles and eliminating the need for them through the formulation of good public policy. Connect-the-dot fails are rooted in the inability to absorb reality no matter how ugly.

The Missing Husband[8]

Like the grief stricken widow searching for her "missing husband", many politicians chase, in futility, the comfort of bygone days. Vladimir Putin sought to resurrect the glory of the Russian Empire because he gave scant consideration to the conditions which gave rise to it and has therefore been unable to generate a new vision.

Putin seeks to recreate his dream with ever more tanks while growth obsessed politicians dream of endless blocks of gleaming office towers, streams of commuting SUVs and fleets of container ships. These fantasies are as destructive to the planet as the attack by Putin's forces has been to Ukraine. They represent the pursuit of worlds we may have once briefly passed through, not ones we can or should permanently remain in.

Failure to connect fundamental causes with their end results makes it difficult to coherently strategize our way to a successful future.

Similar reality-disconnects flourish in regions whose governments' top priority is extending the run of yesteryear into the future. Despite the flush of revenue some activities can produce, they may only profit very narrow interests and result in complex and long term losses for the society as a whole. Only complete lifecycle accounting will provide full perspective.

Big Business Isn't Necessarily Good Business

Over the 20 years of large scale oil sands production, public debt in Alberta has gone from near zero to $100 billion and remediation debt has increased by between $50 billion and $200 billion for a net loss of $150 to $300 billion or $37,500 to $75,000 for each of Alberta's four million residents.

Similarly, Australia has seen its economy focused on housing and resource extraction, in this case substituting coal for the oil sands. This rapid growth has been accompanied by skyrocketing public debt from $60 billion in 2004 to $680 billion in 2020 or $27,000 for each Australian. In terms of environmental costs, the health losses stemming from the coal mining are estimated to be $2.4 billion annually while the remediation costs of the large coal strip-mining operations at $26,000[9] per hectare are a relative bargain compared to those for the oil sands.

Clearly, the oil sands, as distinct from conventional oil and natural gas in Canada, are a net money loser despite the huge top line sales numbers they

---

[8] Based on a segment from the movie "The Nice Guys", www.rottentomatoes.com/m/the_nice_guys
[9] COST-EFFECTIVE REHABILITATION OF MINED LAND IN THE STRIP-COALMINES OF QUEENSLAND BARRY GOLDING.

post. Coal in Australia may well be a net money maker but its complete environmental and social costs have yet to be tallied.

"This system is just not sustainable," said Lucija Muehlenbachs, an economist at the University of Calgary who specializes in the energy industry in a CBC interview.[10] "The promise of this production was that companies would clean up their mess" said Nikki Way, a senior analyst at the Pembina Institute, a clean energy think-tank based in Calgary.

The future does not look bright for the national welfare if these countries continue to make the large footprint, debt-soaked industries of resource exploitation and population growth their core industries. Good national government demands that the full social, fiscal and environmental costs be laid out in a completely transparent manner for all major projects and sectors. Corruption loves opaqueness.

Left Hand, Please Allow Me to Introduce You to Your Right Hand

During the 2018 Ontario election the Conservatives under Doug Ford railed against the high cost of electricity and wasteful curtailment and blamed everything on windmills and solar farms. Ford ran on reducing electricity rates and eliminating inefficiencies, particularly those related to curtailment. Once elected, the Conservatives eliminated many solar farm electricity contracts, cancelled EV incentives and even pulled some EV charging stations from public transit parking lots.

Fast forward to 2021 and we find the Ford government subsidizing hydro consumption with general tax revenues to the tune of $5.6 billion annually[11] while raising kWh rates at the same time. In 2017, Ontario curtailment rates were quite low at about half those of China. And curtailment in a mature renewable energy grid will go to zero given the very large storage capacity of geothermal and an electrified ground fleet. Further, as V2G enabled EVs become widespread, the grid will actually become more efficient. To remain consistent with their election platform though, the Conservatives created the worst-of-all-worlds outcome for Ontario residents. This fiasco was the result of the absence of a grid model, the denial of climate change and the blind pursuit of business-as-usual.

---

[10] https://www.cbc.ca/news/business/alberta-orphan-wells-liability-audit-review-1.5433603
[11] https://energyontario.ca/

The Great Texas White-Out[12] of 2021 is a classic political fail. Close to 130 countries all around the world, with all types of weather regimes have dependable wind generators. Picture yourself standing on the balcony of your 20th floor apartment overlooking the parking lot at 7:30 in the morning after a cold front has moved through. 130 people come out and get in their cars to go to work. 129 people drive away.

The remaining person gets out of their car and, instead of calling a tow truck, takes a bullhorn from the trunk and begins shouting that cars don't work. The media picks it up and before long the internet is on fire with "cars-don't-work" stories and tweets. The story behind the event is that 129 people had the foresight and competence to put anti-freeze in their radiators.

If you don't put anti-freeze in your rad and the engine block cracks during a cold snap, it is not the car's fault.

How Quickly Can We Realistically Convert to Renewables – What Is Involved?

Timeliness matters. A study from Columbia University concluded that early measures could have prevented 84% of the 1 million + American deaths caused by the pandemic. How much pain we experience during the transition to renewable energy will depend largely on how quickly we get underway.

While it would be ideal if we could decarbonize by 2030, as called for in some quarters, it is simply not possible. Factories to produce solar panels, wind turbines, batteries and heat pumps have to be built and staffed and supplies of raw material have to be ramped up to feed them. The complexities of changing energy horses in mid-stream have been hinted at by the production chaos which occurred during the pandemic. Large facilities were sporadically shut down by waves of higher infection rates triggering shortages of key parts with the end effect of reducing production, delaying deliveries and increasing costs.

It isn't difficult to summarize what renewable infrastructure we'll need to construct to be able to arrive in the post-carbon era with our social structure largely intact. We need to electrify. We must reduce energy and material consumption by a very large factor. We need to have a grid with storage infused into it at every possible level. We need to change our building codes to aim for

---

[12] https://www.iea.org/commentaries/severe-power-cuts-in-texas-highlight-energy-security-risks-related-to-extreme-weather-events?utm_campaign=IEA%20newsletters&utm_source=SendGrid&utm_medium=Email

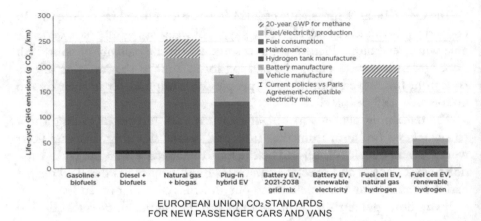

Fig. 11.9 Lifecycle CO_2 impacts of passenger car technologies. (https://theicct.org/publications/global-LCA-passenger-cars-Jul2021)

energy positive buildings and we need to build geothermal storage infrastructure for every region south of the permafrost line.

And as the graph below so clearly illustrates, fine-tuning existing measures, say the gas car fleet with biofuels, or implementing half-measures, such as plug-in hybrids, simply won't work well enough to be effective (Fig. 11.9).

An immediate and strong start in these key areas will put us in the position of being able to get ahead of the problems over a period of 3–4 decades. That means 30+ years of well thought-out and consistent effort with the short term payoff being the somewhat distant rewards of seeing carbon emission levels fall and health and air quality improve. These aren't exactly the glittery visceral pleasures that our large-living society has been brought up to expect.

Immediate actions can include:

- Reassign oil drillers from drilling 3000–12,000 foot deep oil and gas wells to 100–300 ft. geothermal boreholes. The number of feet drilled in 2019 in the US[13] was 250 million feet or 75,000 km. Ground temperatures increase by only an average of 1 °C per 50 m of depth[14] making holes for most uses over 100 m impractical.
- Solar panels need to be installed on every suitable roof surface.
- Stop subsidizing things which are bad and start incentivizing things which are good by building the full cost of processes into the end-user price.

---

[13] https://www.worldoil.com/news/2020/6/25/us-drillers-broke-oil-production-records-in-2019-despite-lowest-rig-count-since-1975

[14] Ground Temperatures G.P. Williams, L.W. Gold.

- Incentivize capital investment to stop building pipelines and start building charging networks and district heating systems.

A cheaper and easier conversion process would involve EVs coming with much smaller batteries say 20 kwh which could be upgraded at any time and a dense grid of fast chargers.

The kinds of things which need to be done and the scale of the undertaking are covered in David MacKay's remarkable TED Talk[15] on Renewable Realities for the UK. Those realities apply to every country. The scale of the task involving wind turbines, solar farms, grid upgrades and fossil fuel de-commissioning projects means that only federal governments are capable of establishing the overall strategy and coordination necessary to see the process through.

Steps Forward

For the majority of countries which truly take the climate crisis seriously, there are many big and small steps which can be made which, collectively, can yield very positive results.

- Korea and Japan will no longer finance fossil fuel projects.
- Many investment funds are 'decarbonizing".
- Rewilding is essentially allowing habitat to restore itself.

  - Italy[16] is rewilding[17] 50 km^2 by "Letting the spaces evolve and grow in their own way will support the creation of five nature corridors that promote both greater biodiversity and capture more carbon in the region, as much as 85,000 tonnes to be exact."
  - The Scottish highlands is perfectly capable of supporting forest but currently they don't because of two factors, sheep and deer, which graze seedlings and prevent the growth of trees. Once the foragers are removed, these protected areas do indeed grow forests.
  - Is it possible Scotland will one day erect a lumber mill and start turning out sustainable plywood? The ancient Picts will be cheering from their graves.

---

[15] https://www.youtube.com/watch?v=E0W1ZZYIV8o
[16] https://www.euronews.com/travel/2021/05/05/what-to-expect-from-the-rewilding-of-italy-s-largest-mountain-range
[17] https://rewildingeurope.com/

– Sweden and Germany have strong silviculture policies which have enabled healthy forests to re-establish themselves. These countries are now largely forest lumber self-sufficient. Many countries ringing the Mediterranean which don't have the policies in place to allow a forest to regenerate after centuries of overgrazing have depleted the soil. Sheep and goats build deserts, not forests.

- The UK Green Industrial Revolution[18] is a £12 billion initiative designed to create 250,000 jobs in clean energy, transport, nature and innovative technologies in the following sectors:

  – Offshore wind
  – Hydrogen power
  – Nuclear power
  – Electric vehicles
  – Public transport
  – Zero-emission planes and ships.
  – Energy efficient homes and buildings
  – Carbon capture
  – Nature
  – Innovation and finance

The EU's Copernicus Climate Change Service "Data in Action" is using modeling to look ahead and enable industry and government adaptation by predicting future climate impacts.[19]

Carbon budget rationing – spend it any way you like but you only get so much – is a concept that falls into the "extreme measures" category. It has been the subject of experiments in countries from Australia to Finland, but can it ever be evenly applied?

Tipping Points Can Signal Rapid Change

Once the infrastructure ducks are lined up, change can happen very quickly. In the late 1800s, horses were the principle means of transporting people and goods in cities. But they presented problems according to Ralph Torrie at The

---

[18] https://www.euronews.com/living/2020/11/18/the-uk-s-green-revolution-plan-is-nowhere-near-enough-say-critics
[19] https://climate.copernicus.eu/data-action

Corporate Knights.[20] A horse required 1 ton of oats and 2 tons of hay annually and there were 140,000 horses in Manhattan. This resulted in an impending crisis of manure and dead horses in the streets.

By the late 1800s, many North American cities used horse carts running on steel rail networks which crisscrossed most downtowns. The key elements of tracks and roads were already in place when electrification occurred, and therefore electrification of city transit systems happened very quickly. Toronto completed electrification of its transit lines in a mere 3 years.

Highways and charging networks (home based) exist throughout the developed world and the transition to EVs can happen as quickly as automobile production capacity can be switched over from ICE production. Instead of the health and odorous issues of manure and carcasses, the existential threat of earth's climate exiting the Goldilocks zone should be enough to spur rapid adoption of EVs. On an individual level, the lure of lower cost for a better product will be enough to make people abandon fossil fuel for electrons as quickly as they abandoned hay and oats for electrons 130 years ago.

What Will Interfere with Countries Transitioning Quickly and Well?

"Successful" can't be interpreted to mean hiccup-free and there will be setbacks but the structural problems which stand in the way of positive action are as follows:

- People
  - People don't like change. People don't like to pay more up front. People don't look very far down the trail.
- Free trade WTO agreements restrict local solutions because they forbid local favouritism and initiatives. However trade agreements may crumble as countries drive hard to create their own domestic medical supply and vaccine production facilities and then extend this to green initiatives.
- Migration "rights". Large scale migration from low ghg emission countries to high ghg emission countries will spell climate failure on a global level.
- Fake solutions abound and without full accounting have the potential to delay and deflect real, productive projects.

  - Carbon capture

---

[20] https://www.corporateknights.com/voices/ralph-torrie/

- Back end loaded announcements
- Biofuels in many cases
- No monitoring
- Proxy metrics - $ instead of carbon

- Investing in and re-structuring of industry. Direct government interference in market and industrial structure is fraught with both real and imagined problems and these will be exploited by vested interests.
- Culture and market change. The super-consumer must become a pariah in the public mind rather than a hero and must be replaced by the parsimonious role model. But it is harder to sell responsibility than SUVs and super-sized servings of French fries.
- The balance between citizen rights and citizen responsibilities will be difficult to define initially. Even the clear-cut issue of wearing a mask in a pandemic became divisive in some regions.
- Power elites who gained their wealth and power from consumption driven markets won't take their loss of privilege lying down. There will be resistance at all levels.

Globalism and Resiliency

Many governments and parties have invested a great deal in the concepts of globalism and free trade which were supposed to result in high levels of efficiency leading to more goods for everyone at lower prices. A huge amount of cheap goods have been produced but in terms of social and personal benefits, the "economic growth lifts all boats" theory simply doesn't float particularly when the lens opens up to include the environment.

Much has been said about the environmental and social impacts of globalism but one aspect of the one-world-market movement that is beyond dispute was the loss of resilience by the "developed" nations brought into focus so sharply by the pandemic. These countries were no longer as developed as they used to be and many could no longer manufacture significant, let alone adequate, amounts of medical critical supplies or renewable energy components.

Had China experienced the kind of outbreak India has, international shipments of supplies and vaccines would have dried up. What of the future for countries which want to de-carbonize but lack the productive capacity to do so?

Disparity

The world's richest man, Jeff Bezos has commissioned a $500 million 417-foot superyacht[21] that requires its own "support yacht". The term "ostentatious consumption" falls short here. Perhaps Bezos will commission a third yacht to add to his yacht fleet to carry members of the press so they can report on what a great life he has.

This is clearly an extreme case of "Wretched Excess" – where resources available exceed worthwhile ideas by a large margin. It is also a tragic failure of leadership in which a wealthy individual, who not only doesn't lead, but sends the completely wrong message to his fellow citizens at a critical time in history.

Currently, 72% of the world's population live in countries facing the double challenge of running a biological resource deficit and generating a less than world-average income. Given that every person needs biological resources for food, water, shelter, clothing, and energy to operate things, global ecological overshoot amplifies the risk of resource insecurity.

The effects of constraints on normal life and disparity have been clear to see in most countries. Where considerable disparity already exists or is rapidly created, angry citizens who already feel cheated by the system may find a change of any kind to be the last straw and spill out into the streets.

The Netherlands is thought of as a nation with very high equality but nevertheless has experienced anti-lockdown riots. Why?[22]

Dr. Jelle van Buuren offers the following analysis in a Euronews interview. "There is a connection and that is distrust in the government, hate against the government, and even more broadly, hate and distrust when it comes to all sorts of societal institutions. Employment precarity and the rise of flexible contracts, in particular, have demoralized young people. Housing, too, has grown increasingly expensive for many. Discontent with the EU has also noticeably begun to increase as has criticism of the country's immigration policies with its multiculturalist society being firmly put under the microscope."

And a Covid-19 curfew was enough of a spark to turn that existing feeling of disenfranchisement into street riots. A paper by Global Footprint Network in the journal Nature Sustainability titled "The Importance of Resource Security for Poverty Eradication" followed the same theme.

---

[21] https://www.cnn.com/2021/05/10/business/jeff-bezos-yacht/index.html
[22] https://www.euronews.com/2021/01/27/angry-citizens-who-hate-the-system-why-the-riots-in-the-netherlands-go-beyond-a-covid-curf

When the tone of government changes from all-is-well and beneficent support to one of demanding some restraint and even winding down once dominant industries, resentment is certain to surface. To counter this and speed the change to productive and sustainable activities, governments have enacted programs by, as the International Energy Agency recommend "putting people at the heart of the energy transition".

The collapse of various industries over decades have given governments some experience with the issues at hand as Albertan and German support plans for their phased-out coal workers show. But instead of just dealing with one or two industries, the transition to sustainability will deal with re-forming the industrial structure itself.

Existing high levels of disparity will endanger or undo any transition process that exacerbates them. Vaxx and mask protests had deeper roots.

Scarcity

As energy becomes more scarce and expensive, so too will most other activities and commodities. Mitigating and adapting to scarcity must be a top national policy priority.

The only things which can reduce the impact of scarcity are technological innovation, conservation, downsizing, recycling and consumption foregone.

Clearly we are losing the scarcity battle as the real cost of commodities has increased as the productivity of our resource bases has decreased. We can see this in the richness of the resources we harvest. Instead of watching oil gushers erupt into the sky from virgin oilfields, we now have to crack rocks deep underground and suck out the oil. Instead of the baskets the first explorers used to haul cod from the Grand Banks, we now have to send trawlers far out to sea with immense nets to get a worthwhile catch of less desirable species.

We have to work harder and use more energy to get less. In the coming decades with the passing of the oil age, we will no longer have access to the cheap energy which made our intense resource harvesting possible.

Long Life Consumables

Scarcity can be partially countered by increasing the useful life of the products we create. Instead of planned obsolescence, aluminum and carbon fibre cars and steel roofs can, along with building structures lasting hundreds of years, reduce our material consumption enormously. And when they finally reach the end of their lives, all things should be nearly 100% recyclable.

The reuse and recovery rate for automobiles in Europe is now about 87%[23] and will go higher. Most EVs, like the BMW i3, now use high percentages of recycled materials. Wood flooring can now be maintained and recycled much more easily with a system developed by Steller[24] hardwood. The system uses no nails, screws or glue and is held in place by extruded channels in a system very easy to install and repair. After 30 or 100 years, just pull the boards up[25] and run them through a sander and re-install. Natural stone can replace asphalt and concrete for walkways giving thousands of years of use.

And of course, geothermal systems have a battery, the earth, which will do its job, millennia in and millennia out, until the end of time.

Most Successful Nations

Internationally, nations have pulled together to successfully tackle global problems such as acid rain and ozone layer depletion which provide some hope the efforts underway to drive decarbonization will successfully stave off a climate disaster.

On a national level, some countries have stepped out in front and posted strong results in specific areas using a variety of different approaches.

Japan has a strategy called "Green Growth[26]" which differentiates between growth in renewables and green sectors rather than general commercial economic growth. GDP is binned in this strategy which provides a roadmap to achieving the goals in different sectors. It forecasts a projected 30–50% increase in electricity demand as fossil fuels are replaced.

The strategy identified 14 industries, such as offshore wind, as well as autos and rechargeable batteries and roadmap for each sector. Under the strategy, the government also provides tax incentives and other support to encourage investment in green technology,

Costa Rica has stopped deforestation and reversed the decline of its natural resources. 75% of its land is once again covered in forest and the tiny country now protects 5% of the world's biodiversity. The title of the story in Euronews was "How did Costa Rica become the Greenest and Happiest Country in the

[23] https://ec.europa.eu/eurostat/statistics-explained/index.php/End-of-life_vehicle_statistics

[24] http://www.stellerinnovations.com/

[25] https://www.youtube.com/watch?v=0jEqHBIL6So

[26] https://abcnews.go.com/International/wireStory/japan-adopts-green-growth-plan-carbon-free-2050-7 4906323?cid=clicksource_4380645_7_heads_posts_headlines_hed

World?" Actually Costa Rica is not the greenest or happiest country in the world but clearly, they are a lot greener and happier than they used to be.

In 1997, Costa Rica developed its "Payment for Environmental Services" program designed to fight poverty and deforestation at the same time. According to Gustavo Segura Sancho, Minister for Tourism, "Had we chosen a model of aggressive growth in mass tourism, then there would have been no conservation of nature."

The Chinese model is one of a competent, responsible technocracy with well-grounded policies focused on the benefit of the people and rigidly enforced. Whether this is paternalistic or authoritarian, the Chinese system is not democratic enough for many people in western societies but results are the metric here. Despite building coal fired generating plants at a hectic pace, China is far outpacing the world on solar and wind development.

The Norwegian North Sea oil development model features a cohesive democratic society where all stakeholders did get a seat at the table and created the best approach based on the interests of the nation and majority of the people. Results: The world's largest national fund worth over $250,000 for every man, woman and child in the country, with next to no legacy costs or public debt. This should be of special interest to Alberta which has pumped a similar level of oil and natural gas but has racked up $100 billion in public debt and many scores of billions of dollars in legacy costs leaving a debt burden of over $50,000 for every Albertan.

Least Successful Nations

Ultimately, national governments are responsible for national integrity which is to say, living up to its commitments to global health and delivering on meeting its ghg emissions targets (Fig. 11.10).

Canada's Broken Promises from the Canadian Club of Rome (CACOR).

Despite pronouncements of purity and leadership, Canada's record is well-known to technocrats and scientists around the world and its word on target commitments is likely now largely ignored. Donald Trump's America performed better in cutting emissions in the period 2016–2019 than did Justin "We're Back" Trudeau's Canada. Australia? Different country, same lobbies, same fails.

Most countries have failed to live up to their carbon reduction targets but it has to be said the majority are now gearing up with clear pathways guided by sophisticated models and detailed monitoring. These are critical

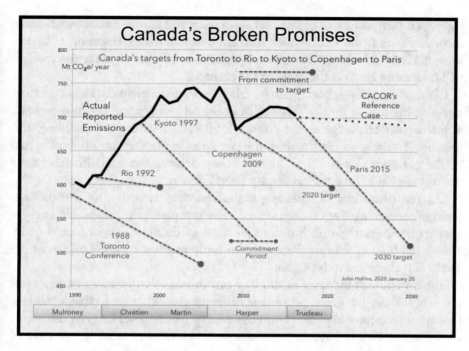

Fig. 11.10 A disgraceful record exposed in one picture. Yet 70% of Canadians think they are world leaders in the fight against climate change

components of success and many countries, are well-placed to make genuine progress towards realistic goals.

Thorny Problems

Assets will be "stranded" when demand for what they are producing falls well below expectations and productive facilities and existing reserves will be left where they are. There are many calls to leave as much oil in the ground as possible to hit avoid exceeding our global carbon budget. Currently, the world has 5 times the fossil fuel it needs to blow through its survivable carbon budget. Continuing as we are now, our budget will be exceeded by 2030.

Obviously, it is best from a global perspective, to leave the dirtiest oil in the ground which means the oil sands in Alberta would be abandoned first with the heavy oil of Venezuela second possibly followed by fracking in many parts of the world.

Walking away from these reserves means stranding more than pipes, buildings, equipment and oil in the ground. It means disassembling the production support infrastructure of plants, transportation and labour forces which make those facilities work. Companies with large capital losses, cessation of operations and no cash flow, will be in no position to pay for cleanup costs which will inevitably land on the plate of the taxpayer.

Chinese coal power plants can be taken down for a great deal less cost than remediating the oil sands. Venezuela's heavy oil sector wind-down presents a different cost challenge as the country's oil infrastructure is crumbling and leaking. If Venezuela can't maintain its facilities while they are producing in a period of favourable oil prices, how will they clean up those facilities and their tailings after the industry has largely closed down?

On the plus side, repatriating manufacturing, remediating the damage amid a comprehensive environmental recovery plan plus renewable energy/geothermal infrastructure buildout all need to be done. These would easily absorb the high quality workers who currently vacuum and boil oil out of the earth. But there has to be a plan.

The investment industry is building carbon sensitivity into its strategy but the atmosphere for green projects is different than current payback models. Green projects have higher upfront costs and payback takes longer.[27] But since all initiatives clearly have to transition to renewables eventually, the risks of not transitioning are very high.

Mark Carney in his "Tragedy of the Horizon" presentation gave an overview of the disconnect between traditional business decision making and that necessary for nations to successfully avoid the worst impacts of climate change.

> "We don't need an army of actuaries to tell us that the catastrophic impacts of climate change will be felt beyond the traditional horizons of most actors – imposing a cost on future generations that the current generation has no direct incentive to fix." ….
> "The horizon for monetary policy extends out to 2–3 years. For financial stability it is a bit longer, but typically only to the outer boundaries of the credit cycle – about a decade. In other words, once climate change becomes a defining issue for financial stability, it may already be too late."

Aside from direct physical impacts, there are many ways climate change can affect financial stability.

• Changes in policy, technology and physical risks could prompt a rapid reassessment of the value of a large range of assets as new costs and new opportunities become apparent.

---

[27] Net-Zero Banking Alliance Eric Usher.

- The speed at which such re-pricing occurs is uncertain and could be decisive for financial stability. There have already been a few high profile examples of jump-to-distress pricing because of shifts in environmental policy or performance.

In other words, the dynamics of the marketplace will be roiled as once bedrock assets are revealed to be on the cusp of becoming marginalized or even stranded. Governments have to be able to produce policy that is well explained and consistent over time to allow investors to make well-considered decisions.

The 85% Transition; How Easy Will It Be?

This section, condensed from "The Renewable Energy Transition, Realities for Canada and the World", outlines some of the mechanics involved in transitioning from a society which gets 85% of its energy from fossil fuels to one that depends on fossil fuels for only 15% of its energy. Past that point it will then take many more decades to fully transition to 100% renewable, zero carbon energy.

In 2017, 85% of the energy for our group of countries came from fossil fuels, which matches the world average. At the point, where we are able to use hydro and natural gas as fast reaction base loads to fill in for variable solar and wind, it will be possible to completely invert the energy mix of the fossil fuel age and rely on renewable energy for 85% of our energy supply. Nuclear energy is left out of this example because, although it could be a stepping stone to 100% renewables, it is not a renewable itself. The 90 + % greenhouse gas reduction which would accompany this shift will be sufficient to meet the Paris 1.5 °C target (Fig. 11.11).

In the chart above, perhaps Americans experience the biggest declines in their energy budgets followed by Canadians, Russians and Australians. But this does not mean certain disaster or deprivation for any of their populations. The US has extremely inefficient transportation and will realize huge cuts in energy consumption by electrifying and modernizing its passenger and freight infrastructure. In Canada's case, replacing almost all oil with natural gas will eliminate the huge amounts of energy currently consumed by the oil sector. This is also true for both Russia and Australia whose largest energy consuming sectors are fossil fuel extraction. As well, both have very large and untapped potential for renewable energy going forward.

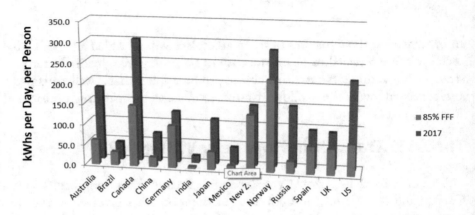

Energy Budgets per Capita 2017 and 85% Fossil Fuel Free

Fig. 11.11 Energy budgets will fall dramatically post-carbon. 85% FFF means 85% fossil fuel free

Steps to Cleaner Energy

Germany

Germany has a large manufacturing sector but this was primarily enabled by the energy from its extensive coal reserves as well as those of Poland. A Germany running on renewables sees its energy budget fall from 127 kWh per day per person in 2017 to 104 kWh in the final stage of the transition. Germany has made a very strong start on renewables but the difficulty it faces is the lack of high levels of hydro and also the common problem of storage.

Like many other nations, Germany has looked to other nations to solve its resource supply problems in the past. For future energy storage, many European nations look to Norway's hydro and dam capacity. But even with its prodigious (for five million people) resources, these can only address a small part of the Europe's issues with its 700 million people.

China

Population growth dramatically decreases the ability of a nation to make a successful transition, whereas a declining population improves that ability via higher ratios of largely fixed renewable resources per person. China, as well as

Japan, can expect to see reductions in their populations of close to a third by 2100, giving them an improving relationship with their renewable base.

China had an energy budget of 71 kWh per capita in 2017 and this would decline to 24 kWh in the 85% formula. However, China has large hydro resources, a falling population and the ability to equip itself with the necessary renewable energy generation infrastructure. These factors indicate it will be able to provide sufficient energy for the government's goal of a semi-sophisticated lifestyle for its people.

China has stated in their Net Zero Plan[28] that their carbon output will decline after 2030 and reach neutrality by 2060. In 2020, built-out their green infrastructure happened at a rate of 5 large 2 mw wind generators and 5 football fields of solar farms every hour.[29] Is this in line with the Paris 1.5 °C target? Debate continues but the scale of China's efforts is impressive.

USA

Despite the political ups and downs in American green energy programs, the country is moving forward in many areas. One notable slippage though has come in the form of the energy consumed per passenger vehicle mile. Already the world's highest over the past several years, transportation consumption has actually gone up while it has declined in virtually every other country. Americans currently have an energy budget of 216 kWh per day which falls to 76 kWh post carbon.

American hydro assets are not impressive, but the areas of southern California and the Southwest offer excellent solar potential, with almost double the capacity factor of installations in higher humidity areas further north and along the east coast. There are large temperate areas in the US which make little in the way of energy demands, yet are capable of providing substantial amounts of year-round renewable energy.

Britain

Without an Empire and compliant states to feed it resources and energy, the UK faces very difficult choices. From the reference 2017 level of 71 kWh per person, consumption slides slightly to 63 but, from this, it likely would have

---

[28] https://www.iea.org/commentaries/china-s-net-zero-ambitions-the-next-five-year-plan-will-be-critical-for-an-accelerated-energy-transition

[29] Just Have a Think www.youtube.com/watch?v=Q4Efc3_b0Mk

to now supply a great deal more of its own food and manufactured goods, greatly reducing the level of "disposable energy". The UK has a very large population on a very small resource base and it will have to be very careful and determined in the choices it makes. The moderate northern climate does make significant energy demands but is not as harsh as that in more northern countries.

Britain has implemented strong energy conservation measures in the past decade and its gross consumption has gone down from 8% since 2005 despite a population increase of 6% over the same period. The country also has gotten off to a strong start on renewables, which currently supply 11% of its total consumption.

David MacKay, in his YouTube TED Talk,[30] implied that the UK needed to make arrangements to be able to use someone else's energy resources, perhaps those of a dry and sun-drenched African nation willing to supply upwards of 25% of Britain's energy requirements. It would seem that, despite large investments in conservation and wind power, it might be prudent for a substantial percentage of the country's population to move to more temperate regions.

In the post-fossil fuel era, this is a consideration many northern nations with modest natural energy supplies might be forced to entertain. Fossil fuels allowed many northern countries to sustain much higher populations but the support systems will look very different post carbon.

Norway

To make countering the very large energy demand of its northern location much easier, Norwegians have more hydroelectric power per person than anyone in the world, and stand to transition to clean energy more easily than any northern country. Their biggest energy sink is that of oil exploration and development, which disappears in the near elimination of oil consumption.

Currently, Norwegians are using 284 kWh a day but, post-oil, they still have a world beating 223 kWh available from hydro to deal with the heavy requirements of their very northern, rugged and transport intensive country.

Too far north for solar panels to be broadly useful, Norway can get by on its hydro, and when they are ready to install it, wind power. Without the export of large amounts of oil, to subsidize the importation of high levels of consumer products, Norway will almost certainly have to increase its

---

[30] https://www.youtube.com/watch?v=E0W1ZZYIV8o

domestic manufacturing which will eat into its disposable energy budget. Although energetically expensive, this will broaden its economic base, and increase their level of resiliency.

Also, Norway has one of the lowest levels of food self-sufficiency in the world. In order to protect themselves from food supply and price shocks, they may have to commit a significant part of their energy generation to food production. What happens if the gulf Stream weakens and the rain patterns change? Then the immense asset of hydro resources turns into a vulnerability.

Canada

Canada is a large northern country with large but diffuse energy flows. The energy required to maintain a high standard of living is among the highest in the world. Canada has significant hydro resources and good wind potential. It has excellent potential for producing solar hot water to energize geothermal storage systems. If Canada continues on a rapid population growth path fueled by mass immigration, it will never be able to maintain its standard of living or food independence. The current government plan is to triple the population of 38 million to 100+ million by 2100 making large cuts to carbon emissions impossible.

Also, if any significant portion of the population moves to the northern portion of the country rather than staying close to the US border, the crunch of higher energy demand and lower energy generation potential would derail sustainability. With its significant hydroelectric capacity, a population in the 20 million range could be supported energetically speaking.

Spain

Spain's daily per capita energy budget drops from 94 to 68 kWh but, given their moderate climate and potential for very large solar and wind energy production, Spanish society has considerable flexibility in determining what level of consumption they feel is necessary and desirable.

Brazil

Energetically, Brazil seems to have it all, with significant hydro capacity, very high solar potential and one of the few large-scale viable sources of biofuel, namely sugarcane. Its current per capita daily budget of 45 kWh declines by formulas used here to 33 kWh but Brazil has the capacity to exceed that level with sufficient investment in solar infrastructure.

Mexico

Mexico has relatively low energy demand given its moderate climate and very strong renewables potential, but despite a moderating rate of growth, its population is still increasing by 1.3% annually, an even faster rate than that of Canada. Oil production has declined 50% since 2005 and natural gas imports are now larger than its own slowly declining production. Mexico spent its oil bonanza quickly and, at the same time, failed to invest in renewables despite its immense potential. It has many arid and desert regions and is approximately 20° further south than Spain, a proven solar powerhouse.

The average Mexican energy budget is currently 43 kWh per day, and by the formula used here will fall to 12 kWh, which is a huge and socially destructive drop. However, given its solar potential, only national will stands in the way of increasing this substantially. Sophisticated societies thrived in Mexico before fossil fuels and they can once again reach great heights on the foundation of modern renewable energy technology.

Cuba, a Real World Shock Transition Experience

Changing from one primary fuel to another and modifying the way life is lived and work is done in a country is a very complex process. Having experience to draw upon is very useful and there has been one "energy shock" event on a national level that provides insight.

The crumbling of the Soviet Union in 1991 caused Cuba's oil subsidy to come to an abrupt end. Cuban agriculture had to convert from mechanized and chemically fertilized methods to organic almost overnight. The population was forced to walk and bike more, and there was a cut in food calories from 3000 to 2000 per day accompanied by a 40% drop in protein. Cubans lost an average of 12 pounds and it took 5 years to bring agricultural production back fully with organic methods. But deaths from cardiovascular disease fell by a third and adult-onset type-2 diabetes by a half, while overall mortality rates went down.

Cubans ate more fruit and vegetables as they could grow themselves. Cuba's energy shock proved to be manageable and even provided some health benefits. But Cuba is a semitropical country. Would a sudden loss of oil be quite so manageable in a northern country, where crops cannot be grown all year long and walking and bike riding long distances is problematic for at least 4 months a year?

The Green Energy Transition Cards Nations Have Been Dealt

The table below puts many of the factors involved in the transition to renewables in one place. It also leaves out a great many. Agricultural potential, fisheries, recycling capability, nuclear and many others would all have a significant impact on the process and the end results. Furthermore, the values in the table are arbitrary and many of the categories are simply value judgements. Energy demand, population size, planned growth and hydro capacity are all based on real data. Hydro capacity is largely fixed as most of the best sites in the world have been dammed.

Energy demand due to a northerly location has a huge influence on the both the need for energy storage. Hydro dams are a large factor in the ability to store high grade energy. The capacity to produce all of one's own energy is also quite easy to quantify. Consumer aspirations and technological capability are purely value judgements and immigration based growth policies can be reversed very quickly.

Whatever the positive and negative surprises in the process, by early in the next century certainly, and very possibly within 40 years, every nation will count itself as a largely renewable energy society.

Different Circumstances Lead to Different Levels of Difficulty in Transitioning (Table 11.1)

Brazil has a relatively low per capita energy demand while that of Russia, Norway and Canada is high. Finland and Germany are approaching population stabilization while China's population will begin to fall in the near future and Japan's is already in decline.

India has a relatively low level of per capita consumption while the USA and Australia are deep into the consumer society class. Per capita wind and solar potential is perhaps highest in Brazil, Spain and Australia while seasonal storage requirements are highest for Norway, Russia and Canada.

Table 11.1 Some of the factors to be considered when planning a path to 100% renewable energy

	Energy demand	Pop growth policy	Consumer aspirations	Tech base	Hydro potential	Energy net Imp/Exp	Solar/wind potential	Storage required	Ease of transition
Brazil	-2	-2	-3	5	3	8	4	-1	12
Spain	-3	2	-1	5	2	3	6	-2	12
New Zealand	-4	-6	-3	6	8	3	5	-3	6
Australia	-4	-8	-8	7	2	8	8	-1	4
Japan	-5	7	-2	8	2	-3	1	-4	4
Norway	-8	-5	-4	7	10	8	3	-8	3
China	-6	5	-1	9	2	-5	2	-4	2
Mexico	-2	-4	0	3	2	0	4	-1	2
France	-5	0	-2	8	2	-2	3	-4	0
Russia	-10	2	-3	6	5	6	4	-10	0
Germany	-6	0	-3	9	2	-3	3	-4	-2
India	-2	-4	-1	4	1	-1	1	-1	-3
Finland	-9	0	-2	8	3	-2	2	-6	-6
Sweden	-8	-6	-3	7	6	2	3	-8	-7
USA	-5	-4	-8	8	3	-2	2	-5	-11
Canada	-10	-8	-8	7	7	8	2	-10	-12
UK	-6	-4	-4	6	2	-8	1	-6	-19

Finally, the full transition to renewables will be easiest for high tech countries, with stable or declining populations, low energy demand and a minimal need for seasonal storage. It is therefore not surprising that Brazil, Spain and New Zealand top the list.

Countries with high storage requirements, rapidly growing populations and high consumption will find the transition more difficult. The UK with a very small hydro capacity, large and growing population and a net dependence on foreign sources for food, energy, commodities and finished goods may have the most difficult time of all. After 5 centuries of living off the resources of other countries, the UK will finally have to become broadly self-sufficient. It remains to be seen how well it will be able to fit its consumer foot into the boot of sustainability.

But the UK has made a strong start in many areas and may overcome its natural deficits to pass better endowed countries which have started late and are yet without strategies.

Hydroelectric power is a huge asset in making the transition to renewables of solar and wind practical. The higher the hydroelectric power capacity per person, the more seamless, with fewer sacrifices the transition will be.

If a nation is dependent on foreign sources of energy, it will be increasing limited in what it can do and will be vulnerable to disruptions in energy supply. The potential of a nation's landscape to produce sufficiently high level of renewable energy per capita will have an immense impact on its ability to prosper as the opportunities to import green energy may well be quite limited. This lays the groundwork for wars of energy "lebensraum" along the lines of most previous wars which were fought for food, water and resources.

Several Large Caveats

- China and its manufacturing output.
  - A great deal of the material wealth enjoyed by people around the world originates in China and was made with coal-based energy. Chinese coal reserves currently have an estimated lifespan of ~35 years. Once China chooses not to be the factory to the world, production of finished goods will, at least partially, flow back to the countries where they are consumed. This process will have a significant impact on both the energy budgets of the Chinese people and those of its former customers. Exports currently consume 11 kWh/day per person of China's energy budget of well under 100 kWh. When coal is eliminated and their energy budget

shrinks dramatically, will China still find devoting a large part of its energy budget to exports to be worthwhile?

- Electricity Bonus
  - Electricity based economies are much more efficient than fossil fuel ones. Therefore gross energy demand can decline very significantly without loss of production.
- Population
  - The bottom line for economic well-being is the amount of disposable energy available to each individual. That determines the work an individual can do whether it be expressed in terms of travel or production of goods or food. The size of the population relative to its energy resource base is critical.
- Natural Gas Bonus
  - By replacing coal and oil with natural gas, carbon dioxide emissions decrease dramatically for the same energy produced. Compared to natural gas, oil (gasoline) has 35% higher emissions and coal 80%. Remediation costs are also likely to be lower.
- Renewable Energy Infrastructure Overbuild
  - The solar and wind generating energy structure has to be sized to not only deliver the power that is needed at the time of use but to generate excess power for storage and use later. The degree of overbuild is related to the amount of storage available, the reaction times of the base loads, (hydro, gas) and their penetration levels (% they make up of the total energy capacity).
- Energy Storage
  - Electrical energy storage requires a great deal of infrastructure investment and it reduces EROIs dramatically BUT geothermal heat storage is extremely cheap. Also, as the ground transportation fleet goes electric, localized electrical battery storage capacity grows to a very substantial level. Make sure the EV you buy has Vehicle 2 Grid (V2G) capability.

Note that all of the above considerations are biophysical and none are based on monetary metrics. Commercial economics, sometimes referred to as "business accounting" in Russia, holds that the larger the economy, the more it will be able to spend on energy and therefore all energy and resource issues can be solved by the continuous growth of the economy. If your country is using monetary metrics and macro-economic tools as the basis for achieving sustainability, their record will resemble the series of miserable failures that Canada's has delivered. (Figure 10.10: Canada's Broken Promises).

The realities faced by different countries in transitioning to renewable energy are not only based on their geography and climate, they are based on the degree to which cheap and abundant fossil fuels subsidized an unsustainable overshoot of their resource bases. There are many factors which have to be included in a nation's strategy to achieve sustainability but unlike issues which revolve around the commercial economy, global warming and resource depletion are threats which simply cannot be solved by throwing more energy and money at them.

From Planning to Implementing

It's reassuring to see figures on paper or computer screen mapping possible paths to the successful transition to green energy and a sustainable and climatically stable world. However, once pen leaves paper and we start to cut steel, we need to understand the scale of the undertaking. It took over 150 years to create the fossil fuel energy networks which have propelled us to the level of consumption we now enjoy.

There are thousands of coal and gas generating plants, nuclear stations and refineries spread across the globe. Almost all of the heating systems in our homes and factories use fossil fuel as do the automotive and freight transport fleets. Presently, these all work. Further, this infrastructure was mostly built using fossil fuels and low cost resources.

We need to replace this infrastructure before the fossil fuel tank runs dry or before the damage we've done to the planet reaches a level fatal to human societies. How much infrastructure we need can only be determined once we understand our energy budget and the biophysical world around us. Asking questions in the language of net energy, EROI and understanding the benefits and costs of converting most of our energy systems to electric will allow us to arrive at the right answers.

Where Does Opposition Come From? - #1 - Follow the Money

Which are the most powerful interests in your country? How directly do they benefit from population growth, increased consumption and asset inflation? The growth industry typically consists of:

- Media
- Banks
- Developers
- Speculators
- Resource exploitation

How much control do they have on the national conversation? The rich and powerful know what generates their wealth and they are loath to accept challenges to their status. They will manipulate policy to structure the market in their favour. Recognize who profits and it is possible to understand who is in control even if their identity is hidden behind several layers of corporate and administrative structures.

Globalism, mass immigration, the oil sands, housing inflation and private banks printing money didn't just happen. They came about from conscious policy manipulations by people who know where their money comes from.

Institutional Corruption: A Rigged Game Can't Produce Good Results

Bureaucracy and business-as-usual repeatedly demonstrate how public policy can falter and go wrong. In an Abbottsford, BC public school, the maintenance staff screwed the windows shut after teachers opened them to improve ventilation during the pandemic. But maintenance then did an air quality test for particulates and chemicals and found no problem, completely ignoring the current issue of Covid-19 and ventilation. Bureaucracy and leaders may just be totally unaware of the problem and stick to routine because they have no guidance to make any changes.

That example of a disconnected bureaucracy is one obstacle to overcome in affecting change but conscious manipulation of policy is quite another.

It is critical to understand which interest groups are in control of national policy. All sides want to be saved by technology but the sustainable society supporters want to use technology to figure out how to achieve stability and growthists want to use it to figure out how to continue growing. In effect, one group is trying to get rid of the tobacco industry and the other is trying to make smoking socially acceptable.

The Powell Memorandum[31] in the USA of 50 years ago and the Century Initiative[32] in Canada now bring to light the differences between being a market player reacting to market forces and actively making market from through media and political manipulation to serve one's own purposes.

The Amazing World of Endless Growth; bringing you the disasters of yesterday, tomorrow!

The Powell Memorandum was designed to bring national policy formation under corporate control, a very few powerful corporations actually, and the Century Initiative is designed to use mass immigration to make the real estate and debt markets extremely profitable for the next 100 years. Of course, the profits come from the pockets of the citizenry. In both cases, control of the media and the national conversation has been essential. This process has delegitimized the mainstream media and shut down broad-based discussion. Consequently every issue from vaccinations to carbon emissions has become polarized. In hard terms in today's polarized politics, a politician only has to look good to the people who voted for him. The only people for whom he has to actually deliver solid results is the donor base which includes media corporations and their contributions of favourable press rather than direct money.

In order to be successful, it is helpful if a politician does a good job but in today's political environment, it is even more important that they look to be doing a good job, no matter what the record would indicate. Bold announcements and photo-ops don't necessarily deliver the performance they might imply. Monitoring, clear goals and good data really do put teeth into transparency and are critical for driving effective policy.

Aging: An Environmental Godsend and a Growthist Nightmare

Population growth is slowing down in most of the world and consequently most countries are experiencing an aging trend as mentioned in Chap. 6. Older people consume less and a declining population means fewer consumers and lower infrastructure demand.

This is a huge problem for developers, speculators, cheap labour employers, media corporations and basically any business that depends on ever-higher

---

[31] https://billmoyers.com/content/the-powell-memo-a-call-to-arms-for-corporations/
[32] https://www.centuryinitiative.ca/

consumption and asset inflation. But for most other people and companies and especially for the environment, the aging of our nations comes at a critical time and will reduce demand for energy and material goods. It all but eliminates the need to build additional housing stock and highways on agricultural land and should signal "peak consumption" in human history.

The final stage of the demographic transition will see populations stabilize and perhaps decline slowly to achieve stability at a much lower level. If combined with strong conservation and re-wilding programs, perhaps the human footprint can fall to a sustainable size gracefully.

Forward looking governments will embrace aging and take advantage of it rather than mounting futile attempts to reverse it with pronatalist or mass immigration policies. To be effective in countering the aging trend, those policies would need to deliver impossibly high rates of population growth forever.

International Migration

International migration continues to be driven by the age old impetus of green fields on the other side of the hill. Desperation drives people to leave their communities and culture behind is search of a better chance at a stable and prosperous life. In the third decade of the 2000s, migration is increasingly driven by the decline in the resource per capita ratio caused by depletion, climate change and population growth and the resulting social chaos in lower income countries.

As well, powerful interests dependent on population growth in developed countries are just as desperate to drive high levels of immigration to their countries because their citizens have chosen not to have as many children.

In probably the largest migration in history, over several hundreds of years, millions of people left Europe for the New Worlds of the Americas, Australia and to a small degree, Africa. Effectively they migrated from high energy and resource demand regions with a greatly depleted resource base to regions with a much healthier resource base and, excepting Canada and the Northern US, lower resource demand.

The pandemic demonstrated that problems are best solved in their place of origin and can't be solved by spreading them around. The indigenous communities on the receiving side of the great European migration is well-placed to bear witness to the damage migration can do to receiving countries.

To their voices could be added the experiences of the low income and middle classes in Australia, Canada and the USA as well as Britain. These people have seen the costs of their housing inflate while their wages decline or stagnate due to the mass immigration policies of their development –obsessed governments over the past 5 decades.

From a global perspective, transporting people from regions of low resource demand to regions of high demand is the most environmentally damaging policy currently in place. The average immigrant to Canada sees their carbon footprint jump by a factor of over 4 and, if their country of origin is in Africa, that factor jumps to 15 times greater stress on the planet.

If 1 billion people immigrated to northern regions, it would have the environmental impact of adding over 10 billion to the globes lower latitudes.

Ideally what should happen to instantly reduce ghg emissions is that people from northern regions should migrate to Spain, Italy, and Morocco and in the Americas, to the area encompassing South Carolina to California and Mexico.

In fact, the large populations currently maintained in some comfort in northern regions are only able to do so because of the abundance and built in storage characteristic of fossil fuels. The renewable energy world, with its lower available energy and storage issues, may be much less accommodating.

National governments have the right and obligation to control immigration in their best national interests and with an eye towards the long-term welfare of the planet. They have a clear responsibility to include the environmental and ghg emissions impacts of population in their planning.

Domestic demand for energy, food and resources has to be balanced against the ability of the country to produce them sustainably. As well, the root causes of migration have to be addressed in the nations in which the problems occur. It is the responsibility of developed nations to help address these issues as Tunisian President Kaïs Saïed, among many others, has repeatedly pointed out.[33]

National Options: Selling Natural Assets or Developing People

Selling off natural assets via mining and unsustainable harvesting practices ultimately leaves the public to deal with the legacy costs. Avoiding paying the full cost for their activities is an extremely lucrative business model for some companies akin to selling a truckload of goods and leaving town before they pay for them. Similarly, population growth generated housing and asset

---

[33] https://www.euronews.com/2021/06/05/no-security-unless-we-eliminate-the-real-causes-of-illegal-immigration-said-tunisia-s-pres

inflation have made fortunes for developers, speculators and finance corporations which never have to pay for the enormous social and environmental costs.

Driven by powerful lobbies, Canadian and Australian governments are following a paved-earth strategy of endless growth and urban sprawl. Unlike scorched-earth, paved-earth can't be remediated and its malls and subdivisions require large amounts of resources to be maintained. The size of the human enterprise has to stop growing but this change flies in the face of the current elites.

A nation with a stable population, healthy environment and a broad based economy with a learning and productive workforce can enjoy high levels of equality and resilience. These are the outcomes when a nation invests in its people. Conversely, a fast growth, resource based country typically has high levels of inequality and housing inflation and high personal and government debt even before the resources play out. It also creates powerful elites for whom continued population growth and resource exploitation are the very blood of life.

Good and bad outcomes don't just happen. There is a clear choice between market building and nation building. Nations can choose to change to invest in the future or continue to hope that the past is the model which will continue to remain valid. Here is an example.

The Canadian government has invested over $4 billion to support the building of the $12.5 billion, 1150 kilometer Trans Mountain pipeline which will transport oil from Alberta overseas through a west coast port. This will allow increased production from the oil sands, the dirtiest oil on the planet. On the other end of this will be increased remediation costs going well into the billions of dollars again out of the taxpayer's wallet.

Contrast this to an alternate decision to spend all of the $12.5 billion on an EV charging network. Manufacturing would be set up to build the systems and those jobs would stay as opposed to pipeline jobs which disappear when the pipeline is finished. Manufacturing the charging components and installing them would build expertise and skills of great use in the future. Building pipelines is in line with making buggy whips. It was good business at one time and it may take talent but there is no future in it.

$12.5 billion buys close to 50,000 charging stations. According to Natural Resources Canada, there were 6007[34] electric vehicle charging stations in Canada in early 2021. Increasing their number by a factor of 9 would more than complete the country's EV infrastructure and allow smaller batteries to

---

[34] https://www.energyhub.org/ev-map-canada/

be installed in a sizable portion of the EV fleet, promoting a faster rollout with lower dollar and material costs.

Instead of building the infrastructure that will be used for centuries, the Canadian and Alberta governments chose to double down on an industry which has lost scores of billions of dollars and has no future in a decarbonizing world.

Nation paving isn't nation building

Structuring the Commercial Market: Political Will and Competence

The following example of the NorthVolt[35] plant in Sweden illustrates the kind of decision making process[36] necessary to restructure an economy and point it in the direction of sustainability.

A reliable supply of batteries is a core requirement in order for a broad-based economy to have a viable automotive/ground fleet sector. Industrial depth is required to assure resilience. Clean energy and high degrees of recyclability are required at every level of production.

The NorthVolt project came about because it was realized that the battery industry in China and Korea is largely coal based so with the 80 kWh of electricity needed to produce a 1 kWh of battery storage there comes a very high carbon load. NorthVolt is based on the hydroelectric power in Sweden and Norway.

The NorthVolt process blends green energy with recycling employing something called a "hydrometalogical process" for a full lifecycle strategy. Instead of using large quantities of energy to separate compounds by burning, a water process (hydromet) is used to capture over 90% of the elements resulting in low energy consumption and a lower raw material and battery cost.

The project is huge and even assuming it works as planned it must have been a stressful exercise for politicians, to involve themselves in this kind of technically advanced effort. Building pipelines, cutting forests and even building automotive assembly plants is relatively simple and familiar ground. Can politicians move out of their comfort zone and build the future rather than the past?

---

[35] https://northvolt.com/about
[36] https://www.youtube.com/watch?v=cOHjIav97HM

Regulations, Standards and Incentives

Government needs to lead on changing consumption patterns using energy, carbon and financial models with multi-decadal time frames to determine overarching priorities which individuals and companies simply can't address.

The federal government has to assure that the jigsaw puzzle of incentives, taxes and standards for conserving energy will be expressed in dollars so they are energetically and environmentally coherent.

Subsidies by themselves typically go on forever whereas incentives are designed to drive change. Incentives can be short term to ease the cost of new technology or they can be disincentives like a carbon tax to make the undesirable activity more expensive. Government has to legislate to encourage innovation and adaptation rather than embedding business-as-usual in its regulations and being the anchor instead of the leader.[37]

Coherent policy won't be arrived at by conflict. One area which needs clear legislative leadership is that of the home rental and large multi-family structure conundrum. How is it possible to increase energy efficiency when the landlord, who controls investment, has no stake in energy efficiency because the tenant pays for energy? Why would the renter make a capital investment?

Governments need to set energy use standards and they need to provide energy efficiency and generation grants using a surcharge on energy used by tenants. In the landlord/renter case, market forces won't deliver progress so the government must lead with standards and the renter will inevitably pay for the upgrades, but in the end, the renter will also be the one benefitting.

Condo and multi-family buildings should have obstacles removed which inhibit upgrades for energy conservation and generation such as EV charging, exterior window blinds and solar pv and hot water panels on the building and property.

The IEA report "Our inclusive energy future"[38] again stresses the point that a society must take the journey to sustainability cohesively and leave no one behind. "As countries seek to advance their shifts to clean energy technologies, the success of these efforts will rest on enabling citizens to benefit from the opportunities and navigate the disruptions. This includes social and economic impacts on individuals and communities, as well as issues of affordability and fairness." The IEA also lists 7 principles[39] for implementing Net Zero policies.

---

[37] https://www.cbc.ca/news/canada/newfoundland-labrador/tiny-home-big-problems-pouch-cove-couple-1.5468181

[38] https://www.iea.org/programmes/our-inclusive-energy-future?utm_campaign=IEA%20newsletters&utm_source=SendGrid&utm_medium=Email

[39] https://www.iea.org/news/seven-key-principles-for-implementing-net-zero

Poorly designed carbon offset and trading can produce very slow emission reductions or, in some cases can give the illusion of reduction while levels actually go up.[40] It depends on whether the intent of the program designers was to produce results or to paint themselves green.

Here is some discussion from an expert in the field.

How efficient is the carbon offset program?

This all depends. There are many arguments for and against the European Union's ETS, which is the world's largest (or at least was), but it is starting to have a real effect now. The cap of the EU ETS in 2050 is very low in line with the targets of the Paris agreement and the cap goes down every year from now until 2050. But the problem in my view is that it is backloaded in the sense that the biggest investments in reducing emissions can be postponed until closer to 2050.

It doesn't give the necessary incentives to make serious investments in mitigation technologies now. It is completely designed for the actors to always take the cheapest mitigation measures first which in commercial economic theory makes perfect sense. The problem is that with the scale of the climate challenge we actually need to make some really large investments in technology now in order to have a chance of meeting the long-term targets.

The ETS is not structured to provide the incentives to do this. There is therefore a risk that even though the cap in 2050 is in line with the Paris agreement, the cap becomes so tight it becomes politically difficult to maintain the reduction in the cap because the costs become very high at some point if all the large investments come closer to 2050.

Essentially this also has to do with the fact that CEOs of the companies that are subject to the ETS answer to short term targets like stock prices and none of them are CEOs by the time we get closer to 2040 or 2050. It is the same thing with politicians so easily signing up to these super ambitious 2050 targets, which are way beyond the electoral cycle, but if you dangle a modest 2025 target in front of them they refuse to sign on...

This article[41] by a World Bank economist makes an economic argument for why we need to start with the most expensive mitigation measures first... which is something the ETSs are not set up to do and is a hard sell politically.. When it comes to carbon offsets there are other challenges more to do with the integrity of the offsets and their global climate effect. The issue with for instance the Clean Development Mechanism (CDM), which is the UNFCCC offset scheme, was that it was essentially a mechanism to make it cheaper for the signatories of the Kyoto protocol (the developed countries) to meet the targets they committed to. So Norway (a Kyoto Protocol country) would buy offsets or credits from a renewables project in an African

---

[40] https://www.cbc.ca/news/opinion/opinion-carbon-offsets-1.5951395

[41] www-wds.worldbank.org/external/default/WDSContentServer/WDSP/IB/2011/09/21/00015834 9_20110921094422/Rendered/PDF/WPS5803.pdf

country (not a Kyoto party) which they could use to fulfill their obligations under the Protocol.

> *There were a lot of good things about this. It made it cheaper to reach the targets. The theory was that this would make the developed countries willing to take on more ambitious targets. This may very well have been the case when the Kyoto protocol was negotiated. It also, at the time, introduced the concept of GHG mitigation measures in a lot of developing countries. It helped put climate on the political agenda in these countries. It did also generate investments in the developing world.*
>
> *The problem was that it essentially made it possible for developed countries to meet their targets without reducing their own emissions."*

For offsets, there is definitely an issue with green washing but a business network in Norway recently issued a Guide to avoid Greenwashing which can be applied in any country.

Green-Washing Rules for Companies

Most corporations want to promote healthy communities and their owners and executives can read a global temperature chart as well as anyone. Companies have a great deal to consider regarding the nebulous world of green claims. Here is a summary of suggestions from the Green Washing site.[42] The advice basically boils down to be sure to make your actions bigger than your words.

> *Greenwashers talk loudly about sustainability without contributing to a better world. Responsible companies work hard to create shared values for themselves and society.*

1. **Be honest and accountable and be careful when using words like green, sustainable, recyclable, eco- and fair, without both explaining and documenting what your company has done.**
2. **Make sure that your company's sustainability efforts are not limited to your communications and marketing departments.**
3. **Avoid talking about the importance of sustainability, nature, the climate and ethical trade, if your company has not made serious efforts on these issues yourselves.**
4. **Do not under-communicate your company's own emissions and negative impacts on the climate, nature and human lives.**

---

[42] https://gronnvasking.no/en/home/

5. Be careful using a big share of the marketing budget on small measures that do not affect your company's footprint significantly.
6. Avoid buying a clean conscience through climate quotas or by letting others clean up ocean plastic.
7. Use established labelling, or work towards the establishment of good labeling mechanisms in your industry if it is lacking today.
8. Be careful using terms such as "better for the climate, nature, and the environment".
9. "Cherry Picking" from the UN sustainable development goals can lead you astray.
10. Donations and sponsorships are great, but not a proof that you are working on sustainability issues.

National Conversation

The health of a robust national conversation is the responsibility of the national government. It must assure that clear goals are always before the public and that its strategies and models are accessible and transparent. All stakeholders must be represented and heard.

National governments must distinguish between activities which simply transfer wealth or inflate valuations and those which actually create real wealth. Identifying the real economy and the biophysical realities which underpin it are crucial to the formation of effective policies. Decision making must allow the public to vote on issues rather than for brands or hollow promises.

If the national conversation degenerates into obfuscation and polarization, the country will begin to register real-world fails. The slide to irrelevance of media corporations has been caused by their failure to listen to the voices of either the general public or science in their efforts to promote their own interests and those of their major advertizers.

The financial interests of a media corporation and its advertizers are very different from and sometimes diametrically opposed to those of the general public. Media company sources of revenue are local and dependent on market size as well as on the structure of the market. If it is growing there will be big splashy new home, car and finance ads. Large subdivisions generate full page new home ads and even entire sections.

This structure of ads generates the largest source of income for newspaper companies and, coupled with increasing circulation means media corps are hugely dependent on a growing population. Picture a stable or slowly declining population generating no large new home ads, no increase in circulation, very little debt and much lower new car sales. Can media corporations survive when 1/8 page ads for "Bob's Kitchen Renovations" replace multi-page ads for luxury condos? Whatever their editorial slant, all media corporations have the same financial interests.

The old media adage of "if it bleeds, it leads" has been replaced by the theme "grow or die" baked into all news and editorial content. Somehow, responsible governments need to be able to rise above the control media corporations have over the national conversation and lead with well-grounded, forward looking and inclusive legislation.

Clear National Leadership

National governments have to work with all levels on all issues because, at the end of the day, the buck stops there. Without goals and strategy, public programs become ineffective and piecemeal as Bolsonaro in Brazil and Trump in the USA demonstrated during the pandemic. Strong efforts were made by many local and state governments and all front line medical workers. They struggled to make the best of a chaotic policy environment caused by polarized and disruptive leadership. As a result, hundreds of thousands of people died unnecessarily.

> **Bhutan Index of Happiness - Gross National Happiness**
>
> PM Jigmi Thinley – "Bhutan is not a country which has attained GNH. Like most developing countries we are struggling with the challenge of fulfilling the basic needs of our people.
>
> What separates us from most others however, is that we have made happiness - the foundation of human needs - as the goal of social change."

National governments must take back control for national welfare and initiatives from a globalized system which makes no allowance for public or environmental concerns. They have to maintain a full set of social, fiscal and environmental books and they are responsible for the integrity of the data in the system. Data quality is as important to a society as the integrity of the voting system is to a democracy and perhaps even more so.

Developed nations have achieved more than high enough levels of material consumption to assure personal satisfaction as well as health and education infrastructure. Now they must turn their attention to sustainability and resilience to avoid entering the downside of the social progress cycle which started in the early 1700s.

The approach taken by many countries towards Covid-19 was to "flatten the curve" which meant enacting just enough restrictions to keep hospital services from becoming over-stretched or collapsing. This turned into an exercise of futility as the complexities of Covid-19 and social and economic dynamics produced repeated lockdowns and chaos in the goods and service sectors.

The pandemic lesson is that it is impossible to successfully fine tune a response to a complex and ever changing problem. The approach has to be one of forthright and consistent action to grind through to a quick and clear conclusion of the threat.

National governments must be prepared to change gears and commit whatever resources are necessary to put their countries beyond the reach of the major biophysical threats which are now on our doorstep. Maintaining normalcy and business-as-usual can no longer be the prime directive. In all of this, governments must establish price and incentive mechanisms which will allow individuals and companies to make energy and resource-rational decisions.

Old news is really no news when it comes to conflict over scarce and diminishing resources.[43] The earliest discovery of relatively large scale battles or ongoing raids of eradication took place in Sudan 13,400 years ago. It appears different groups had to withdraw into the centre of a diminishing resource base as climate change and probably their own over-consumption depleted the landscape. Conflict was inevitable.

National governments have to assure they create the biophysical and social environment for peace and stability rather than depending on pieces of paper or large arsenals to prevent hostilities when competition for resources among desperate groups makes conflict a certainty.

---

[43] https://www.standard.co.uk/news/uk/sudan-british-museum-nile-b937630.html

12

How to Invest for Minimum Disruption and Maximum Benefit

Transitioning to renewable energy and reduced consumption will bring individuals back into contact with the natural systems and energy flows we became disconnected from in the fossil fuel era. This is a whole new world where there will be a myriad of opportunities to invest to reduce your cost of living and increase your own security. These range from systems on your own property to participation in neighbourhood initiatives. Social cohesion will be increased through building community gardens and their energy equivalents – geothermal, district heating, micro-grids and recycling programs.

Citizens can drive active transportation with pedestrian and cyclist friendly community designs offering more space and better health outcomes and greater energy independence.

At some point, a family's clean energy aspirations have to be fused with personal and business budgets based in dollars. If the government has done its job properly, dollar incentives for investment in conservation, generation and storage will closely adhere to biophysical budget parameters. This will produce an optimum real-world result assuring that a good financial decision will be a good energy/emissions reduction decision. However, looking at government track records in dealing with the pandemic, it is best if each homeowner does as much of their own homework as possible.

A good strategy recognizes that everything is done in steps in a process that can be staged over years or even decades (Fig. 12.1).

J. E. Meyer, *The Post-Pandemic World*, https://doi.org/10.1007/978-3-030-91782-1_12

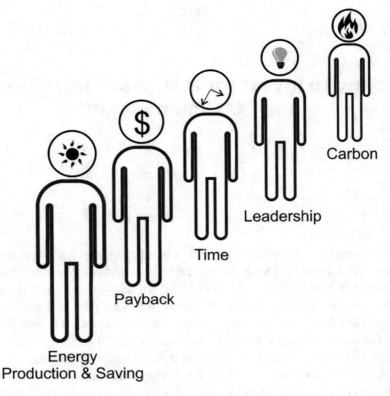

Fig. 12.1 Energy based decisions require a different perspective than simple money based decisions

Your Priorities: Putting Your Mental Ducks in a Row

You are now of several minds when it comes to consumption, conservation and investment. You have an energy brain and a carbon brain as well as a money brain. How are you going to prioritize these? How much money and how much time will you invest in making your aspirations a reality?

Properly done, one process leads to another and produces a synergistic system which produces greater results than the sum of its parts. Installing solar panels on the roof and putting an EV in the driveway puts a huge dent in carbon emissions as well as reducing living costs substantially.

Such a house is likely to provide greater energy security as well as a healthier living environment. At this relatively early stage of the renewable energy transition, these two quick and simple upgrades will draw many questions from friends and neighbours.

Leadership by example is leadership in its most powerful form. Being able to offer firsthand knowledge and numbers to those around you, whose circumstances are close to your own, is the best formula for breaking down resistance to change. As familiarity displaces fear of the unknown, people around you will begin to undertake their own carbon reduction initiatives.

Your Time

It is possible to make many of the transition steps with a minimum of time commitment depending on the financial resources available. Those with very deep pockets can spend as little as a few meetings with one-stop-shopping consultants, write a cheque and walk away and come back several weeks or months later to a massively lower carbon lifestyle. Nothing is ever quite that simple but the range of time commitment for many of the projects which need to be undertaken can vary tremendously depending on owner interest and financial capacity.

The time/results trade-off is different for each step towards carbon neutrality. Replacing your ICE vehicle with an efficient EV involves the same amount of time as buying any other type of car plus the installation of a home charger and that will likely be arranged by the dealer. So for a few hours of effort which was going to be spent anyway on replacing an older vehicle, an individual can take a very large chunk out of their carbon footprint. And as electricity grids around the world decarbonize by replacing coal with natural gas and renewables, the purchase of an EV will deliver increasing carbon reduction dividends going forward.

Contrast this with the time and effort to crawl into an attic and perhaps drill into outside walls to blow in insulation. It is necessary but requires more time and certainly much more effort and likely delivers lower carbon reductions and financial payback than an EV. However, an EV won't make your house more comfortable to live in whereas improved insulation and sealing definitely will.

Your Money

Changing personal infrastructure from fossil fuel based to electricity based will introduce different ripples in personal finance. Maintenance costs will change as will insurance rates and so may property valuations. Unlike the earliest days of solar PV and battery backups, the insurance industry, whose business is risk management, now has a great deal of experience in evaluating these relatively new systems.

It is hard to make generalizations but it is fairly safe to say that replacing an oil furnace or woodstove with a geothermal heat pump will result in lower insurance costs. EVs have lower maintenance costs overall and going with an electrically based home heating system will mean fewer visits from repair and maintenance teams. Installing a large solar PV array will incur increased weather risks. Susceptibility to hail damage increases but potential loss from storage and use of explosive fossil fuels is removed along with their carbon monoxide threats. EVs have slightly lower rates than ICE vehicles; ask your agent for quotes.

Surplus electricity production can be sold into the grid or "stored" there in the form of credits for use at a later time. With fossil fuels, energy systems were constantly "money out" but once homes and individuals begin to generate and store energy, money flows can be two-way.

Investments in energy conservation and green generation are likely to appreciate as energy becomes more expensive and carbon penalties trend higher.

Self-Interest Versus Responsibility

Throwing oneself on one's sword in the name of the greater good is noble but unnecessary given the state and maturity of most green energy technologies.

Everyone, particularly older generations such as Boomers, who have made the fewest sacrifices in building their countries yet have eaten the lion's share of the resource banquet, have an obligation to show leadership in the movement to reduce consumption.

Have we irreversibly bought into the commercial message that our comfort is all that matters in our societies? Do we still really believe the economics dictum that if everyone just acts in their own self-interest all the time this will guide the markets to deliver a successful resolution of all social and environmental ills?

Such tortured twisting of early economic theory to justify growthist agendas simply doesn't hold water in today's world of stressed environments and disintegrating social cohesion. Markets failed the Covid-19 stress test and they will fail just as badly on resource scarcity and climate change mitigation.

Clearly we all have a responsibility to change our habits and reduce our consumption to the best of our ability. It may be too late to be able to hand our children "a better rice field" than the one we inherited but we can at least head in the right direction and leave behind better tools and a more aware culture.

Note that according to the International Energy Agency, electric vehicles are one of the few technologies on track under the Sustainable Development Scenario. Conversely, carbon emissions fell across all sectors in developed nations in 2020 with one exception – SUVs which consume 20% more energy than the average vehicle. In the USA, this is 30% and, like guns sales, sales of SUVs are at least partially driven by fear. In this road going arms race, weight and bulk equals firepower and armour.

Ultimately, everyone shares the same interest when it comes to planetary health but the tragedy of the commons mechanism ensures there is conflict between the short term benefit of some and the long term benefit of all.

Resilience Costs Short Term Gratification

Resilience is a form of social stability insurance covering flood, drought, fire, famine, financial disasters and power interruptions. The premiums are facilities which are underutilized most of the time, lower personal consumption and higher financial costs.

The benefits are far more control over energy and vital resources with greater price stability to prevent losses and instability driven by natural asset decline. Unresilient societies fracture rather than bend.

Awareness

Awareness of the carbon content of fuel enables a gradual modification of behaviour and buying decisions. The simplest way for an individual to reduce their carbon footprint is to replace all of their fossil fuel devices with those powered by electricity. Different grids have radically different carbon emissions per kiloWatt hour depending on how much of their generation is coal and natural gas based but all grids are gradually replacing coal plants with renewables.

Simply changing habits and lifestyle to avoid using any energy at all is, of course, the ideal but there is only so far one can walk or bike. How low is it practical to reduce the temperature of a house in the winter? Energy use is inevitable but the lower the carbon content of that energy the better.

Carbon Consciousness[1]

- 1 litre of gasoline = 650 g carbon, 2.3 kg of CO_2
- I gallon (US) of gasoline = 5.5 pounds or 2.5 kg and 20 pounds or 9 kilos of CO_2
- 1 cu ft. of ng = 0.12 pounds or .05 kg
- 1 cu metre of ng = 1.75 kg
- Quebec and Norway Grid 1 kWh = 35 g/kWh
- China Grid 1 kWh = 700 g/kWh
- USA grid 1 kWh = 420 g/kWh

More Tools and Fewer Toys

Picking the right tool for the job does not mean buying the biggest tool one can afford. Being electrified makes most anything more efficient than the gas version but not necessarily efficient enough. Right sizing the tool is the key to optimizing efficiency and minimizing cost. Trucks are great tools but terrible personal transports from environmental and cost points of view. Trucks were designed to be tools but have become a fashion accessory for many. In the much more rational, less marketing driven sustainable society, we'll only buy tools we need.

To this point in the EV market, range has been everything. It was the prime concern in the earliest Nissan Leafs as owners struggled to go over 120 km on a charge. But as the market and buyers have matured, range is of far less concern now given the 400+ km some EVs are capable of. This coupled with increasingly fast charging rates and a rapidly growing charger network has largely eliminated range as a constraint for most people.

Now buyers know their own needs well enough to be able to select a vehicle which can do the job at hand; in other words a Honda E instead of a Tesla S long range. Range paranoia is but a fading memory in today's quickly evolving car market with its much better informed buyers.

Instead of buying, insuring and maintaining what you only need very occasionally, consider renting and investing your money in cutting costs and increasing energy production. With large houses and their double garages overflowing into storage units with things we will never use again, we need to declare a war on STUFF and get our collective hoarding gene under control.

---

[1] https://www.eia.gov/environment/emissions/co2_vol_mass.php

Realistic Life Spans

How long is the expected life of a well-treated EV? Estimates now range from 15 years to 30 years. A full decade after EVs first went mainstream, it is apparent the life span of EV batteries is far longer than anyone guessed in 2011.

Even when they cease to be practical for a vehicle, they can be repurposed for other uses as they don't stop working, their capacity just drops. Solar panels were forecast to have a lifespan of 25 years but now it is apparent that their output at that point will likely exceed 80% of their original capacity. Why scrap them at 80%?

Of course, the lifespan of geothermal systems is strictly dependent on the life of the pipes. Even if these begin to leak at some point decades down the trail, the cost of repair will be the cost of replacing the pipes, not re-drilling the entire system. Electric motors? Remember Ferdinand Porsche's first car which was built with the technology and materials of the late 1800s and yet ran after 108 years in storage.

Strategy and Scale: Leaving Links Open to the New Grid

The new grid – the living energy organism – will continue to grow and evolve. When making energy upgrades, be sure to allow flexible links to the grid in order to be able to take advantage of any new developments either on your end or elsewhere in the system.

The new grid – or should we say "family of grids" - will have the potential to generate, store and distribute electricity and heat to and from any connection point whether that point is deemed to be either a net consumer or a net producer of energy.

Home owners should allow for a larger electrical panel box, EV chargers and for eventual geothermal piping entry into the house.

Solar PV and heat, conservation and geothermal storage will become more valuable over time. EVs will decline in cost but their payback is the fastest of all green initiatives right now. ICE vehicle demand will likely go into an increasingly steep decline by 2026 as will their re-sale values.

To put battery electric storage into the context of seasonal storage, consider the scale required to use Tesla PowerWalls, to get through a long northern winter. To early 2021, 200,000 PowerWalls with 13 kWh of storage have been manufactured. If the average wintertime energy consumption for a home in

Scandinavia, Russia, Canada or the northern US was an optimistic 50 kWh a day, it would take 4 PowerWalls to maintain its heat levels for one day. It would take all of the PowerWalls in the world to provide power for only 52,000 northern homes for 1 day in the winter.

This indicates how vital geothermal storage is for northern countries. After 6 years of PowerWall production, (note that production did double in the past year), the cumulative total production of PowerWalls would provide power for 0.4% of Canadian homes for 1 day in the winter at a cost of $2 billion not including installation.

Taking the extreme case of having no supplementary energy production from rooftop solar, wind or hydro for the whole winter of 120 days, the cost of batteries needed to heat 100% of Canadian homes would be $60 trillion. Every 25 years. In 2021 the Canadian GDP was $1.8 trillion. These figures scream *"GEOTHERMAL STORAGE!"* as both an individual and national priority.

How to Prosper in a "No-Growth" Environment

"No growth" is generally portrayed as a "stagnant" or declining society in the mainstream media which is substantially dependent on advertising revenues from companies which absolutely require continuous growth in population and consumption for their existence. But for most businesses and individuals, the higher per capita incomes and lower overall costs accompanying the cessation of simple growth will deliver benefits far outweighing the costs of the transition.

In a "no growth" economy, learning, improvements and progress will still take place and only total material and energy consumption will peak and decline. Who benefits and who does not in a no-growth environment?:

- Business revenues rise in these sectors:

    - Local energy companies
    - Local manufacturers
    - Local retail
    - Local agriculture/food producers
    - Technology leaders
    - Smart, small-scale housing builders (see Drake Landing, Whisper Valley, Springfield Meadows[2])

---

[2] https://www.youtube.com/watch?v=ovytcaA82bA

- Revenues fall in these sectors:

  - Subdivision and commercial expansion development
  - Speculation
  - Large media corporations
  - Debt institutions
  - Cheap labour employers
  - Fossil fuel producers

- Government expenditures rise in these ministries:

  - Environmental remediation
  - Training/relocation and placement
  - Education
  - Urban health and nature overhauls
  - Industry wind-down transitions
  - Green energy and conservation incentives and research
  - Foreign aid/green incentives
  - Infrastructure maintenance and upgrade

- They fall in these ministries:

  - Oil and gas subsidies
  - Immigration
  - Welfare and unemployment
  - Infrastructure expansion

What is the best investment? Young people. Give them a strong grounding in energy and biological systems and let them learn by doing.

For reference, the future of oil demand and pricing is anything but clear. Will there be enough to meet 2050 demand or will we be awash in it? Will demand be curtailed by emergency climate restrictions? Is investment high enough in oil development to assure stable supply? Rystad Energy says maybe not.[3] In any case, the fossil fuel future is likely to be far more chaotic than the renewable energy future.

---

[3] https://www.rystadenergy.com/newsevents/news/press-releases/the-world-will-not-have-enough-oil-to-meet-demand-through-2050-unless-exploration-accelerates/

Renewables will be produced on a much more localized level than oil and natural gas and will be less subject to international supply issues or financial upheaval.

Home and Vehicle Investment

Individuals building a new house or adding an addition, might consider what could be termed a "Smart Skin" approach where the entire surface of the building[4] can vary itself between energy collection and insulation. This is too extreme for most people currently but this technology is certain to gain more widespread adoption as costs drop and building standards begin to embrace energy harvesting.

Homes currently heating with electrical baseboards, propane or oil will find heat pumps to be the best upgrade option with geothermal, as opposed to air-source, being the ideal long term choice. Those currently using the lowest cost heat source, natural gas, will find replacing it with geothermal is initially much more expensive. Even with a rooftop solar PV array the financial payback will be much longer but the carbon benefit is both large and immediate.

Those with an accessible yard or parking area of over 500 sq. ft. (50 m²) available for a geothermal borehole system can consider that option. Below are a number of major investments which will dramatically lower carbon emissions but which have a wide range of financial payback times (Table 12.1).

The above payback times are completely dependent on where one starts.

In terms of security, investments in an electric vehicle with Vehicle to Grid (V2G) capability, geothermal heat storage and solar panels on the roof will substantially eliminate damage and possibly any inconvenience due to grid failure for a period of days.

During an outage, your neighbours will be coming around to charge their cell phones, which is a good thing in most communities. If you drive your V2G EV to another house to run their heating or fridge or sump pumps for a short while, make sure the panel connection to the grid is switched off, which is to say it is in "island mode". If it isn't, the power will go into the grid with enough juice to possibly injure repair crews or people in other houses.

There is a galaxy of proven technology available for investment and paying it forward, (and for most boomers) paying it back for all of the thoughtless consumption we've indulged in through our lives.

---

[4] https://www.cbc.ca/news/science/bipv-solar-1.6044485

Table 12.1 Residential heating system upgrades energy, carbon and $ impacts

Project	Additional dollar cost	Annual $ saving	Energy saving	Carbon saving	$ payback time
1. EV	+$0–$10,000	$2,500	A	A+	3 years
2. Heat pump air	+$8,000	$500	B	B+	3 years
3. Heat pump ground	+$20,000	$1,400	A+	A+	8 years
4. 10 kW Solar Panels	+$30,000	$2,000	A+	A+	8 years
5. 20 m² of solar hot water panels	+$12,000	$800	B+	A	10 years
6. On-demand electric water heater	+$4,000	$200	C	A	20 years

1. Replacing the purchase of a sports sedan or SUV with a Tesla 3, VW ID3 or Ioniq 5, Nissan Leaf, etc. Results get better the smaller the replacement EV is
2. Replacing electric baseboards, propane or oil with air source heat pump
3. Replacing systems above and adding ground source heat pump
4. This is a straight add-on to existing house systems and produces a huge drop in energy costs if the house runs on a heat pump. Also, fill up your EV for "free". If you are pumping excess energy into the grid they will likely charge a ~50% fee for taking your energy, "storing" it and sending it back when you need it
5. For northern climates, and a house with borehole geothermal, this will boost the COP enormously and cut down electricity costs
6. On demand water heaters, whether natural gas or electric/geothermal, can reduce energy use considerably because they eliminate the tank storage of hot water along with the losses of heat that occurs. In the summer, there is much less heat generated in the house that the air conditioning has to negate

Timeframe matters and people with a 5 year financial horizon who are considering a new car should simply go out and buy an EV today and please make it smaller than the vehicle you have now. On every level, from financial to energy to carbon, an EV makes sense right now.

Many people, by virtue of renting or condominium ownership may not pay for energy directly or, if they do, may not have the option of changing energy systems or making large capital investments in their building. For those people, a cooperative approach is necessary to include the building owners in a strategy to both reduce energy costs and carbon emissions (Fig. 12.2).

Community Investment

Life is local and communities are the core of all sustainable living projects. Communities must lay down the information foundation needed to upgrade the skills of their residents and coordinate progress on reducing energy and

We are local

Impact
- Most affected by what happens to house/property neighbourhood, district, town/city
- Further from home less impact
- Impact of local actions on local environment
- Cumulative local effects have global impact

Fig. 12.2 People live in a localized world. (Canadian Club of Rome Presentation Don Cowan 2 June A Strategy for Sharing World Environmental Data & Information https://www.youtube.com/watch?v=iZEm2Ca8e3Q)

material demand while building out clean energy production and storage capacity.

This takes investment of time and money to create initiatives such as:

- A community energy budget
- Educating the young and decision makers – living labs
- Trees, recycling, solar farms, recharging stations, recycling, active transportation
- District heating
- Local suppliers, not just farmers market, a broad spectrum of producers
- Technology and small business incubators, access to expert opinion, training programs, apprenticeships, casual job coordination
- Local knowledge base[5,6]

The above relate to existing communities but, in a few rare instances, there is the opportunity to create a community from scratch. Instead of going through the many steps of upgrading their existing homes and waging campaigns to

---

[5] https://myperthhuron.ca/
[6] http://comap.ca/fwis/

bring their communities forward, Ready-Made community solutions are actually available now in a very few locations.[7]

Marketing takes a new twist away from salacious consumption to appealing to the high minded and practical buyer with the slogan "Sustainable is now Attainable" employed by the Whisper Valley housing development. There is nothing new about the development; it is just that the developers are packaging a practical, low-cost community with all of the latest technology which is:

"A Complete Solution to Save Natural and Financial Resources"
"Our EcoSmart Solution (ESS) combines an innovative geothermal infrastructure with additional energy resources to create an integrated technology package. Together, these elements work to generate power, reduce usage, eliminate noise and cut your utility bills."

- The geothermal based master planned communities in Whisper Valley[8] Austin, Texas are the first of their kind in the USA but may well be the start of a trend. When completed the 2000 acres (800 ha) Whisper Valley will consist of:
- 5000 houses
- 2500 apartments
- 2 million sq. ft. of commercial space (light manufacturing??)
- up to 30,000 residents

This development is built around a Geogrid system taking advantage of the 23 °C Arizona ground temperature and promises low or zero energy bills and very low carbon emissions. It is essentially a scaled up and updated version of the Drake Landing development in Okotoks, Alberta which has over 10 years of proven performance.

The driving technology for the Whisper Valley[9] development comes from EcoSmart Homes[10] whose mission is "designing and delivering an innovative geothermal infrastructure combined with a comprehensive suite of distributed energy resources to enable developers and builders to create energy efficient communities and provide new homeowners affordable, comfortable, and sustainable living."

That, in a nutshell, is the common objective of everyone who sets out on the path of reducing their footprint and increasing the healthfulness and

---

[7] https://www.youtube.com/watch?v=F07RNxIAm9Q
[8] https://whispervalleyaustin.com/
[9] https://whispervalleyaustin.com/the-importance-of-nanogrids-in-low-carbon-residential-communities/
[10] https://ecosmartsolution.com/

Whisper Valley Basics

- Energy efficiency measures such as ultra-efficient appliances and smart thermostats dramatically reduce energy demand.
- With geothermal, home heating and cooling energy demand is 75% lower
- With rooftop solar panel array, net consumption from grid is likely eliminated
- All homes wired to take an EV charger
- All homes get solar pv array
- Flat rate utility bills are based on the size of the lot because larger lots have larger heat pumps and *they put more energy into the Geogrid!*
- Austin electricity provider buys surplus solar panel output at 60% of retail
- EV ownership is 3× national rate
- Geothermal boreholes are 350 feet deep
- From authour's point of view, owners should overbuild solar pv capacity and drive for free as well as being energy positive. The cost of adding extra panels is minimal.
- Current US federal tax incentives allow between $6,000 and $15,000 (or 26% of value) write-offs for pv and battery storage systems.

resilience of their community. Your low carbon footprint life, complete with district heating, geothermal storage and solar energy collection, will start when you first turn the key to the door of your new home.

If these developments were popping up all through existing towns and cities, they would make carbon downsizing extremely simple. They may well become more common as innovative builders see a new market opportunity. However, any cooperative planning a new development or re-development would do well to take advantage of the templates which now exist in Alberta, Texas and the UK.

A Tesla solar development, SunHouse[11] at Easton Park, is exclusively solar based and does not include geothermal district heating or storage at this time. It therefore misses the point that societies run on heat almost as much as they run on electricity. SunHouse does not provide a complete package and it falls short of the Whisper Valley effort. Elon Musk needs to take his electrical blinders off and open up to the entire energy landscape.

---

[11] https://www.statesman.com/story/business/2021/07/14/austin-neighborhood-fueled-tesla-solar-battery-tech/7943383002/

Lead, Follow or Get Out of the Way

Electrification and renewable energy are happening: the question is what is your timetable. Short of moving into a new geothermal district heating community with a new EV waiting in the garage, most people will have to pick their spots to change their home and tools over to electric.

As familiarity with electric cars and larger recreational vehicles grows, acceptance may come faster than anyone could have predicted several years ago. Converting to electric, once thought of as a sacrifice and a curiosity, is now more and more regarded as an upgrade with price being the only obstacle. Once running costs are included though, money saved is now just another strong reason to step away from fossil fuels.

Perhaps journalist Justin Rowlatt of the BBC has gotten it right in his article[12] "Why electric cars will take over sooner than you think". If so, readers should be astute enough to take the implications of this speed of this trend into account in their own personal planning.

The graph below should look familiar to those who followed the early days of the pandemic. Exponential growth looks the same on a graph whether cars or Covid-19 cases are being represented (Fig. 12.3).

The fossil fuel industry is facing a virtual pandemic of EVs with the same exponential growth curve of Covid-19. It isn't really a big problem for them now but 9 years of 40% annual growth would leave EVs with 85% of the new car market and the oil industry with a huge and growing hole in demand for their product. The BloombergNEF group look at the EV penetration of the ground fleet predicts a 70% market share[13] for electric by 2040 which unfortunately still leaves a very large number of carbon vehicles still on the road. For the individual, in the case of electrification and conservation, going with the flow makes sense from all points of view.

Just because the prevailing wind is electric mobility and home heating and cooling, it doesn't mean that future won't include unfortunate events. Large objects, from hail to aircraft parts to tree branches during a windstorm do fall from the sky (Fig. 12.4).

Energy systems may change but risks will never disappear.

---

[12] https://www.bbc.com/news/business-57253947
[13] https://about.bnef.com/electric-vehicle-outlook/

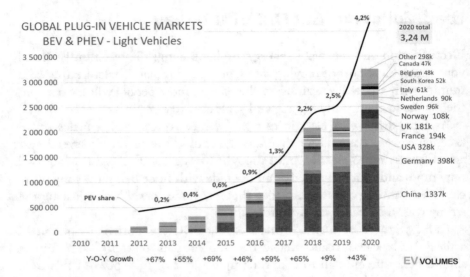

Fig. 12.3 EV sales represent a positive exponential trend from ev-volumes.com

Fig. 12.4 Solar panels are resilient but only to a point

Coherent Investment Strategy

1. Buy an EV. Now.
2. then install solar PV
3. then install geothermal storage
4. then install solar hot water

Along the way, cut energy consumption by doing all of the small things which really do add up. Storm doors, insulation, external rolldown blinds on large windows and patio doors, biking, walking, buying locally produced food and goods and making sure everything you buy has a long life span and is highly recyclable.

In the world of fossil fuels, every device is completely separate from every other device since all power comes from a can of gas. In the electric world, mowers and chain saws will be charged by output from the house electrical circuit which may get most of its power from rooftop solar panels. These tools may use the same batteries which are swapped around as needed. The solar panels can also charge the car which itself can also charge the batteries of the smaller tools for remote use while the car can power the entire house during short term power outages.

Investing in a Back to Physical Reality Future

If you sat down and formulated an action checklist for the next pandemic, what would it look like? If you went through the same exercise imagining that governments had suddenly gotten serious about climate change and implemented emergency measures appropriate with the level of the threat, what would you wish you had done to prepare ahead of time?

An investment strategy of downsized consumption with overbuilt power generation is a formula that provides better preparation for any problems which might come along from any point of the compass. Like a vaccine, this strategy won't grant complete immunity to any threat but it will offer increased protection and will minimize major damage. But whether threatened by bio-physical upheaval or a massive systems hack by malevolent individuals, the more secure your personal energy supplies are, the more resilient you will be.

This Is What Is Coming

What event will trigger your own personal "Pearl Harbour Moment", the point at which you realize urgent action is required? Pearl Harbour occurred on December 7, 1941 and it caused a divided and drifting USA to transform into a unified and determined nation by December 8. So focused and committed were Americans that many young men who were deemed unfit to serve in the armed forces committed suicide out of shame and frustration.

2021 finds the USA and many other nations seemingly locked in a downward spiral of division and decay with national policy mechanisms largely unresponsive. Individuals must step forward to break down the barriers to action and conversation (Fig. 12.5).

The above graphic from the IPCC July, 2021 Summary for Policy Makers illustrates the world we are entering as our planet's temperatures accelerate into ranges never before experienced by humans. It is hard to imagine that we will be able to hold temperature increases to 2 °C.

The complex impacts of temperatures on weather patterns, ocean currents, floods, fires and droughts will play out whether or not you and your community are ready for them. It is much better to prepare before they strike than to attempt to re-build and retrench as repeated climatic, social and environmental surprises occur. The quality of your life is going to be greatly influenced by how well prepared you are to ride out these challenges.

In a pandemic, the ideal is to prevent infection with masks, lockdowns, tracking and vaccines but as far as Covid-19 is concerned we are well past the point of avoiding its impacts. The same holds true of climate change. We are now infected and we need to both reduce the viral load and prepare for the onset of symptoms. Climate change is here and rolling.

Paper Versus Hardware

Governments may well feel they have no other choice than to continue to print trillions of dollars to keep the commercial economy ticking over. The flood of money over the past 40 years has inflated the value of assets, notably real estate and stock markets. For the small number of people who have ridden this wave of paper wealth, the necessity of conserving energy or building energy generation into their lifestyle may seem unnecessary.

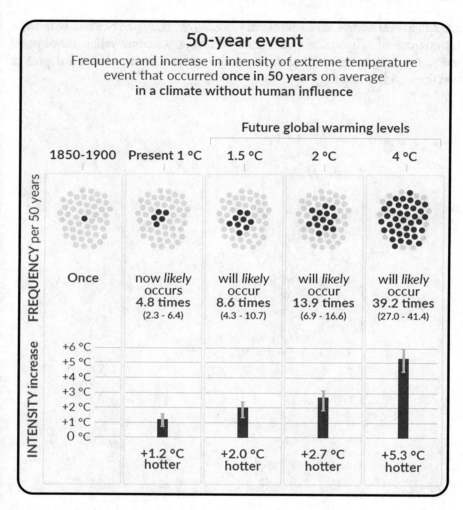

Fig. 12.5 IPCC 50 year event forecasts appear to be quite conservative in the summer of 2021

Possibly paper printing will never come to an end and the gains in "asset enhancement" will simply continue. If history is a reliable predictor of such events though, it seems unlikely that countries will be able to perpetually spin apparent prosperity from money be it fiat currency, crypto or some other baseless claim on real wealth.

Even if the double digit returns of the stock market and real estate speculation occupies most of a person's attention, a little diversification into real world economics wouldn't go amiss. For those who have taken a hard look at the ephemeral nature of our financial systems, the transparency and real world foundation of biophysical economics will be a welcome relief. Biophysical metrics will serve as a sound basis for investment for an elevated degree of certainty and security.

13

How to Drive Change

Lead by Example: Passive Leadership

Working models hold a great deal more sway over passionately espoused theories when it comes to convincing people to make changes and invest in new processes. Sets of numbers recorded from actual events form a solid base for planning by removing the fear of something new. Leading by example builds instant credibility because there is nothing more valuable in a discussion of options than practical experience and real world data.

This is to say that the best way to convince other people to change is to buy an electric car, install solar panels or convert to a geothermal heating/cooling system and then tell other people about your experience. These can be individuals or politicians but whoever your audience might be, they will pay much more attention to someone who actually has done what they are advocating others do.

We have passed the point where any of the major steps advocated in this book represent great unknowns. If you have questions, ask for support; there is an extensive body of technical expertise available to deal with questions associated with any technology mentioned here.

These technologies are supported by enthusiastic and accessible experts who are very willing to help and who may well be quite close by and have firsthand knowledge of local conditions. There is a massive amount of information at all levels available on the internet and it is as current as electronic data exchange can make it.

Storming into an editorial office/politicians office and demanding change is going to have no impact unless you do so with a few thousand friends and

J. E. Meyer, *The Post-Pandemic World*, https://doi.org/10.1007/978-3-030-91782-1_13

even then all you might succeed in doing is generating headlines and hardening attitudes. Asking questions which politicians feel obligated to answer when you already have the answer is a good way to start a dialog which can lead to positive improvements.

National policy change can only come through a change in attitudes and broadened perceptions.

At one time, the development of the Alberta oil sands was widely regarded as good business. But several decades and hundreds of billions of dollars of debt and legacy costs later, the oils sands (as distinct from conventional oil and natural gas) no longer look like good business and have never looked healthy from an environmental perspective. Quite the opposite is true as every politician and screen sensation looking for attention can point to the oil sands as the great environmental evil. They aren't wrong but they aren't telling the whole story either.

Urban sprawl is still portrayed widely in the media as a sign of economic health. In reality it is good business only for very narrow interests as discussed previously and, despite its clean appearance, subdivisions are dirtier in the long run than the oil sands. Once the oil sands are exploited their residue sits largely dormant. It may contain a nasty cocktail of potentially seeping chemicals but it will emit no more carbon.

Subdivisions on the other hand, require constant inputs of energy and material to sustain them. The houses themselves and the expanding population they shelter in Canada emit more CO_2 than do the oil sands. Every country still committed to population growth may not have their own oil sands but they do have expanding urban areas pushing up raw material and energy demand. This urban expansion, featured in glossy sales brochures and proudly touted by strutting politicians, will be eventually be seen as the resource and energy folly they really are; just as the oil sands once were (Fig. 13.1).

Yet urban sprawl is heralded as the true sign of economic health by the media and politicians. "My town is bigger than your town" to them means it is better even if it does have higher taxes and a lower quality of life.

Rather than judge how something might look good on the outside, isn't it better to figure out how it is actually going to work and what the long term effects are? Miss America of 2020, Camille Schrier[1] who appeared on stage in a lab coat instead of a bikini, called the focus on science instead of swimsuits 'freeing'.

---

[1] https://www.foxnews.com/entertainment/miss-america-2020-camille-schrier-science

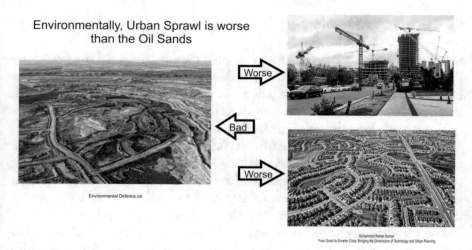

Fig. 13.1 Urban sprawl may look like progress to politicians and media but they are a rolling environmental disaster

We need to free the national conversation of hyperbole and replace it with science and the perspective of a complete view of our society and its place in the environment. If Camille Shrier can win the Miss America contest while pitching good science rather than flaunting a plunging neckline, surely society is ready to hear well-grounded policy recommendations from leaders who eschew the polarized world of half-truths and hysteria.

That is where individuals with firsthand knowledge can play a part, by preparing the ground for a rational national conversation with real people speaking about their actual experiences.

Why base a country's decision making on science rather than commercial interests? Science is self-correcting whereas corruption is self-reinforcing. Commercially based policy can't focus on interests other than those of the current elite and it can't react to biophysical changes and threats.

If the Miss America contest can drop the swimsuit event and elect a woman who conducted a science experiment on stage as her talent demonstration, countries can elect biophysically informed leaders.

Reality cheques written by Mother Nature don't bounce.

Green New Deals

Green new deals are designed to lay down a clear pathway for the decarbonization of a country's economy. The authors of these proposals recognize that both economic and social stress will occur in the transition process. The clean energy infrastructure and consumption reduction initiatives can be considered the hardware side while human impacts form the social side.

In order for the critical infrastructure adjustments to be made there will be economic and social disruption resulting in some groups experiencing more pain than others. Even now, most developed countries face equality challenges and the social side of the green initiatives is focused on maintaining social cohesion sufficient to allow the hardware transition to succeed.

For many countries, such as Japan and particularly those in Europe, the social safety net is highly developed and green new deals are aimed at maintaining the integrity of existing programs through fine tuning. In the USA though, the social safety net is patchy and weak and the American New Green Deal is designed to both drive the clean energy transition and fix the great inequalities which exist in that country.

This is certainly a much greater challenge than most developed countries face. Even in Canada, Australia and the UK, where equality levels have been falling precipitously for decades and divisions have been growing, basic social service programs are in place and able to address the disruption the fossil fuel to renewable energy conversion will involve. In the USA, sufficiently effective programs must first be put in place to prevent further abrasion of the social fabric.

Hence the American Green New Deal harkens back to the New Deal of the 1930s when America was trying to pull out of a deep depression and prevent large numbers of vulnerable citizens from being crushed. Infrastructure building was one way the government chose to put people back to work and give them a financial base. Although the infrastructure was certainly needed, the social side of the initiative was the prime driver.

The situation is now reversed with the green part of the equation being the critical motivator and the new deal portion acting to both stabilize and upgrade social programs. The similarities between GND initiatives and the intended long term pandemic responses of successful nations are striking.

US Green New Deal

- Convert fossil fuel jobs to energy production and conservation

- Convert US industry to producing green energy products
- Upgrade US homes and businesses
- Research support
- Overhaul tax system to promote green energy and conservation and penalize fossil fuel consumption
- Make US less dependent on foreign suppliers and technology
- Social cohesion through a fairer tax system and living minimum wage

This is new ground for the globalized English speaking countries but it effectively is how many European and progressive Asian countries have already structured their economies. Taken in its component parts the New Green Deal is not by any means radical, but it is a radical change from the way the USA, and to a lesser extent, Australia, Canada and the UK have been conducting their affairs.

Comparable Long Term Pandemic Strategy

- overbuild vaccine production capacity
- overbuild testing capacity
- overbuild emergency medical infrastructure
- establish contact tracing infrastructure
- control over border
- build domestic PPE industry
- social cohesion

Building a national strategy around the goal of transitioning to clean energy is a necessity and should be supported, regardless of teething problems such a policy will have. It is critical to get a start on dealing with a clear threat such as climate change.

Individual Leadership

As can be seen from the graphic below,[2] what individuals choose to do about their own personal home and transportation energy consumption and carbon footprint can have a huge impact on the drive to avoid a disastrous climate change. Remember there are eight billion and counting of us and this is how

---

[2] https://www.youtube.com/watch?v=N8aGV3Z8dOA

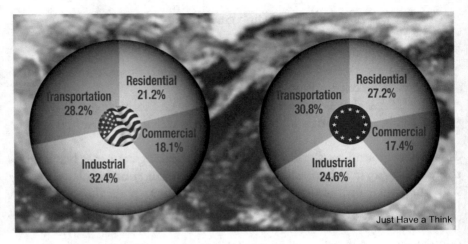

Fig. 13.2   US and European energy consumption patterns

the higher consumption nations spend their energy. Few people realize how much energy they use on a daily basis (Fig. 13.2).

Martyrs Need Not Apply

Individuals spouting theories are only slightly more effective at motivating change than those advocating what amounts to martyrdom in the name of achieving sustainability. In medieval Europe, living in an unheated monastery on a mountainside and chanting prayers for 6 h a day may actually have been an improvement over the lives many endured in squalid poverty.

But in our very comfortable and mobile twenty-first century world, suggesting that we all need to live like monks will not garner a great deal of support. We do clearly need to live smaller and increase our awareness of the world around us though. Once that shift has started, perhaps it will be easier to frame major adjustments to comfort and convenience levels in a rational discussion. A smaller footprint need not mean privation, it can mean "healthier" which is an argument most people will listen to.

Support Your Local Leader

Personal initiatives are critical but collective voices are heard more clearly by politicians and bureaucracies. It may not be necessary to build an advocacy campaign from scratch as there are many like-minded people in every community. Finding these people and possibly the campaigns they have already initiated will provide a structured outlet for your social change energy as well as being a more efficient one than solitary efforts. Still, don't be afraid to tilt at windmills.

Why Institutions Are Change Resistant: Active Leadership

Making changes in your own life and consumption patterns can easily lead to the desire to drive change in the greater community. This can be very productive and fulfilling but it also presents challenges as those resistant to change will confront people trying to promote a new Business-as-Usual.

People are resistant to change but they are less resistant to improvement. "Better" sells more easily than "different". While "bigger" is easier to promote than "better" to cash flow dependent policy makers, most people prefer an improved quality of life to more cars on more congested streets.

How to get involved, maintain your sanity and achieve results:

- Inform the debate
- How to change your planet. What if Greta doesn't answer when you call?
  - You'll have to do it yourself. Channel your inner Greta.
- Don't expect everyone to like you. People who won't like you include:

  - People who won't adapt to anything to the point of refusing to wear a mask in an Ebola ward
  - People who can't make decisions because they don't want to offend anyone whether it is Ebola deniers, rights activists or people with a financial stake in business-as-usual.
  - People who make a living from the things which need to change.
  - People who enjoy and have built their lives on the way things are done now.

    - Which includes most of us

- Call out Greenwash

The end of growth and the beginning of sustainability means shifting the ground under managers' feet. They will have to deal with:

- Stranded hard assets
- Stranded personnel
- Stranded systems which have to be re-thought
- Terms they have never used before
- Infrastructure they have never worked with

Time and a well-considered rollout will reduce these issues from the crisis level to the problem level. The later change is left, the faster changes will have to be implemented and the costlier and less effective it will be.

Denial

Denial of climate change is becoming increasingly rare but if you encounter anyone who does not comprehend the implications of the graph below, it is best to go around them as they will never change their mind (Fig. 13.3).

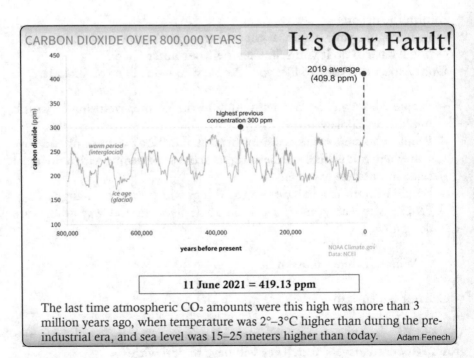

Fig. 13.3 Atmospheric carbon: if this graph doesn't alarm you, nothing will

Exposing Greenwash

People and projects professed to be green but are anything but, abound. Full lifecycle accounting done in kWh and carbon emissions is new to virtually everyone but it is critical to avoid false starts and wasted effort.

- Explain energy rational and irrational examples:
  - Solar panels in the Arctic produce some electricity but less than ¼ of those in the US Southwest. The EROI of these systems is extremely low even before storage is added.
- Ethanol EROI of 1:1 does not make for green gasoline

Canada's Justin Trudeau has projected an image of environmental leadership while President Trump proudly assumed the mantle of an environmental Darth Vader. Image is one thing and reality can be quite another. In fact, in a comparison of CO_2 emissions performance, Trump's USA handily beat Trudeau's Canada over the period 2017–2020. Canada's emissions went up by 3.4% while those of the USA went up 0.6%.

In the case of Donald Trump it was an accident due to all of the state and local initiatives beyond his control plus the switch away from coal to natural gas. In Trudeau's case, emissions went up as a direct result of his deliberate expansion of oil sands production and the high and increasing levels of immigration.

In many cases, no matter how ugly it may appear, the greenest solution is the least bad one. Incinerating unrecyclable garbage, regardless of the real time emissions, may be better than burying the garbage leaving future generations to deal with the consequences.

Mother Nature doesn't hand out participation trophies to those who fail to live sustainably.

Make Sure that Policy Is Guided by Real Numbers Representing the Real Problem

- Use net energy and EROI when dealing with energy
- Include all legacy costs when dealing with dollars
- Well-being is not easily definable but people are real and need to be represented by more than dollars. What do people want most?

Flattening the wrong curve happens. When countries failed to seal their borders and allowed Covid-19 into their general populations the policy then became one of controlling it to prevent cases from overwhelming hospital capacity. This was called "flattening the curve".

Given the dynamic nature of Covid-19 and the inability of segments of the population to absorb the concepts behind health based restrictions, "flattening the curve" would better have been termed "riding the dragon". The wily Covid-19 got loose repeatedly.

A better policy would have been to "flatten the curve" to the point where track-and-tracing programs were reliable. This would have resulted in far more stringent restrictions early in the pandemic since the ability to track and trace Covid-19 outbreaks is far lower than the ability to treat them. These effective programs would have allowed far earlier opening and no large waves. A thorough vaccine program would then have allowed a high degree of re-opening on a permanent basis.

Of course, this can be viewed as hindsight. But it shouldn't be because epidemics and pandemics have occurred throughout human history and into the modern era. We should always have been prepared with the technical infrastructure in place and the policies to roll-out when the first indications of a well-known problem became evident.

Going Broke on Growth

For a few short centuries, simple growth, in the form of population and consumption increases, built well-being, towns and nations. It is now safe to say that very few regions on this planet or their occupants would benefit from more people and more demand. In the middle 2020's virtually every resource base on the planet is already being heavily utilized, and many are well into various levels of depletion.

More growth in demand will lead to more short term cash flow but real costs will increase as the biophysical, and eventually the fiscal, viability of growing regions decline. Do your elected representatives know the difference between what built the prosperity of the region and what is now needed to maintain the prosperity of its citizens?

Differentiating between progress and development:

• Development means building more houses and shopping malls to accommodate an increasing population – it is a metric of the debt/developer interests and is the near perfect negative environmental barometer.

- Progress means increasing investment in people and building per capita incomes, equality and quality of life – it is an indicator of the health of the broad society and its environment.

Besides shifting middle class prosperity from the developed nations to Asian countries, globalism has had the global impacts of higher GHG emissions and increasing plastic waste in the ocean. Covid-19 laid bare the vulnerabilities of long shipping distances, fragile supply structures and reduced worldwide resilience to most biophysical threats. Globalism may have resulted in the lowest front-end costs for consumer goods but it is a global race to the bottom both environmentally and socially. Where do your politicians stand on it? On resilience?

"Endless Growth" is a commercial economic mindset. It isn't present in physics or earth science and Mother Nature has certainly never supported it.

How to Exit a Bad Industry

Individuals, corporations and governments are very reluctant to walk away from working business models regardless of what their specific downsides might be. Shutting down a business is that much harder in regions with little else.

Countries like Germany, Japan and China have broad economic bases and can absorb the displaced workforce and costs associated with closing down a coal, tobacco or asbestos type industry.

Germany[3] and Alberta[4] have enacted progressive policies to deal with the shutdown of their coal industries. These can serve as references for anyone picking their way through the labyrinth of issues surrounding the shift of industry to clean energy and sustainable practices.

---

[3] https://www.abc.net.au/news/2020-02-18/australia-climate-how-germany-is-closing-down-its-coal-industry/11902884

[4] https://www.alberta.ca/support-for-coal-workers.aspx

Making Sure the Table Is Round and All the Chairs Are Occupied

The pandemic saw governmental policy boards, normally stacked with developers and members of the finance industry, re-populated with medical experts, demographers and scientists. In addition, the voices of social equality were clearly heard as lockdown inflicted tremendous financial pressure on those least able to bear it.

So glaringly stark was the threat of Covid-19 that governments the world over saw the need to rapidly abandon their commercial economic blinders and adopt completely different policy priorities recommended by a broad-based and expert group of advisors. This dramatic shift in advisors and policy took place over a matter of days to weeks.

Climate change and a depleting planet have been occurring over decades. When governments finally turn their attention to dealing with these issues in a substantive way, they will need a similarly broad based council of experts to steer towards the new goals. Instead of minimizing the impact of a virus, the goal will be to transition away from the fossil fuels which built the prosperity we all enjoy and to reduce our consumption to levels the planet can live with. Social stress will likely be far greater than that experienced during the pandemic, making the inclusion of lower income and vulnerable industry groups critical to the maintenance of social stability.

Many international groups are working on the question of transitioning smoothly. The IEA has established a commission headed by Prime Minister Mette Frederiksen of Denmark, which will examine how to make sure that people are at the centre of clean energy transitions around the world.

"Our Inclusive Energy Future: The Global Commission on People-Centred Clean Energy Transitions"[5] sees the issue in these terms: "As countries seek to speed up the shift to clean energy technologies, part of their success will rest on enabling citizens to benefit from the opportunities they create but also navigate the potential disruptions. This includes social and economic impacts on individuals and communities, as well as issues of affordability and fairness."

The growth lobby won't have as many chairs as at the table as they are used to and will almost certainly resist the loss of influence and diversion of focus from their priority of GDP growth.

---

[5] https://www.iea.org/programmes/our-inclusive-energy-future?utm_campaign=IEA%20
newsletters&utm_source=SendGrid&utm_medium=Email

Conflict of Interests

There are no social or environmental problems which population growth won't exacerbate.

There are very few commercial interests which will not benefit from population growth.

Who is paying the messenger?

Simple growth has conflicted with the interests of the great majority of people in developed economies for the past several decades. Equality levels have plummeted along with job quality while the prospects for their children have eroded. A very narrow section of the population has benefited tremendously. Call them the 3% or the 1%, those involved in finance and asset inflation have profited immensely from simple growth. Growth, which for several centuries lifted all boats, is now a zero sum game with winners and losers. The nation-building growth of the 1600s to the 1950s, through which most prospered by way of increased production and higher incomes, has been replaced by growth in the commercial economy where transfers and inflation boosted cash flow while job quality retreated.

This shift has not been covered by the media corporations whose ad revenue is dependent on the banks, developers and mass consumer retailers which thrive in this commercial bubble.

This makes it extremely hard to have a rational, information based national conversation. When a politician's major donors and the media corporations have one interest, it is hard to hear any other voices.

In the 1950s, some politicians were able to speak up against the growing power of arms industry made powerful during the Second World War.

"In the councils of government, we must guard against the acquisition of unwarranted influence, whether sought or unsought, by the military-industrial complex. The potential for the disastrous rise of misplaced power exists, and will persist.

Now this conjunction of an immense military establishment and a large arms industry is new in the American experience. The total influence—economic, political, even spiritual—is felt in every city, every Statehouse, every office of the Federal government. We recognize the imperative need for this development. Yet, we must not fail to comprehend its grave implications. Our toil, resources, and livelihood are all involved. So is the very structure of our society."

President and General Dwight D Eisenhower Farewell address 1961

"I'd let you talk more, but you're
not as interesting as me."

Fig. 13.4 The media does not find science or public opinion interesting or profitable

This speech applies to all elites and particularly to our current developer/debt economy but what voices are heard today when the elites own society's printing press? (Fig. 13.4).Guilded by "enlightenment", "greenness" and apparent sophistication, the media's core message is one of the need for unrelenting growth. The national conversation has been reduced to a one way monologue from media corporations. With a huge amount of skin in the growth game, they tailor the national conversation to suit their own interests. They are now a "grow or die" industry.

"State capture" is the term applied to the process by which small elites determine national goals rather than governments. This requires control of the media and the national conversation.

Corruption Is Socially Corrosive

When the interests of the elites in power diverge from those of the majority of the population, disaffection occurs. There is a loss of trust in institutions. Media corporations lie so often it is assumed by some that everything they say is a lie. This has the unfortunate effect of negating the effectiveness of genuinely competent coverage of vital issues and reducing medical and scientific professionals to the same level of trust as commercial economists and politicians.

Declining equality levels and the quality of life for most people has occurring against the backdrop of soaring wealth for a very few. It has resulted in a general disengagement from public discourse and the fracturing into many polarized and non-communicating shards of the national conversation.

The national conversation is the bedrock of a democracy and it is built from a base of scientifically sound and commonly acknowledged data; a common worldview. It must prioritize a discussion of what people actually want rather than what politicians and their backers want to give them.

Americans once believed in their own abilities to get ahead. They had a can-do attitude. Now the majority of people no longer believe working hard will lead to a better life[6]. The 2020 Edelman Trust Barometer reports that despite strong economic performance, a majority of respondents in every developed market do not believe they will be better off in 5 years' time. Money wins elections and often buys favourable press which is critical to winning most elections. Donald Trump managed to find a way to use any press coverage to win an election. Ask who your politicians' largest donors are. For municipal politicians, it is likely the largest donors are developers. This has to be noted front and centre.

Governments have to be able to measure and represent public welfare to assure public order. One of America's founding fathers, James Madison, warned in 1822 that "A popular Government, without popular information, or the means of acquiring it, is but a Prologue to a Farce or a Tragedy; or, perhaps both."

He was imploring that an informed and involved citizenry was necessary for democracy to thrive. Today we see a distant citizenry picking their own facts from whichever source appeals to them. Currently, a well-informed, cohesive national conversation is effectively dead in many democracies.

"It's a lot easier to believe something if your job depends on believing it."

Your Own Personal Corruption index:

---

[6] https://www.abc.net.au/triplej/programs/hack/2020-edelman-trust-barometer-shows-growing-sense-of-inequality/11883788

- The process is corrupt or trending corrupt if:

  - politicians are unable to state clear goals

    - or are unable to define metrics

  - There is no mathematical model or pathway towards the goals
  - Benefits, in any form, flow from stakeholders to decision makers
  - There is no environmental or social well-being impact assessment built into plans.

Asking Questions Which Demand Answers

Why do some questions demand answers? Because the people being asked know they should know the answers and so do those around them. Ask one or two specific questions. Short succinct questions can be answered while long treatise won't be absorbed or answered specifically.

Specific questions on goals and means of measuring progress towards them should at least start a conversation. Follow up when necessary. Exposing that no goals or strategy exist is not the end, it is a start. Clarity on the lack of goals and metrics can spark action from local activist groups.

Standards of performance must be created. They exist for financial issues in the form of accounting statements, forecasts, budgets etc. No agency would make a move or could exist without them. Why aren't they in place for sustainability questions? Modelling capacity exists to determine renewable energy infrastructure. Is your government using them?

Hold your local and federal governments to standards being achieved by leading jurisdictions. Rest assured that "business-as-usual" will be aggressively defended by some interests. But has BAU been good for the majority and will it continue to be?

Have you Been Sierra Clubbed?

Politicians who say one thing and do another are commonplace. Advocacy groups which charge at the towers of power but later mute or dilute their initiatives are harder to spot but they definitely exist. It is easier for vested interests to go inside activist groups and deflect initiatives aimed at their cash cow than it is to confront them in public.

- Does one or more member drone on endlessly at meetings and demand more specifics, higher standards to the point where any initiative grind to a halt or is completely watered down?
- Long winded rambling "orators" might be doing more than just listening to themselves talk; they might be rigging meetings so that no one else can talk or extending meetings to the point of exhaustion so nothing is accomplished.
- Has your prime mandate been diverted?
- Has a large donor influenced policy or arranged to have their choices put on the board of directors?
- Volunteer organizations by definition have a hard time getting things done but if one or two individuals are making it noticeably more difficult by virtue of red-herrings, procedural morass, diversion, fear etc., then a discussion of core values and action plans is needed. This may become pointed.

A 2004 *Los Angeles Times* article[7] broke the story about David Gelbaum's $100+ million donation to the Sierra Club on the condition that they not address immigration. The money was accepted by the club with the clear knowledge that they would have to abandon any work on a population policy. This meant essentially eliminating any possibility the club could either develop or present an effective strategy for American sustainability.

> "I did tell Carl Pope (Executive Director) in 1994 or 1995 that if they ever came out anti-immigration, they would never get a dollar from me."

In 1979, while a grad student at UBC, camping out to save money on rent, Leon Kolankiewicz used an unexpected, modest scholarship to pay $750 for life Sierra Club membership. Now he regrets giving the money he could have spent on meals to the Club, to which he also devoted considerable time. He now is feeling "swindled" by an "organization that supports, in effect, endless population growth just like the very corporate interests they pretend to oppose" (Fig. 13.5). Those with deep interests in endless growth know the best way to hobble dissent and discussion by activist groups is to lead those groups themselves. Controlling the conversation is critical to be able to assure the continuation of endless-growth policies.

---

[7] https://www.latimes.com/archives/la-xpm-2004-oct-27-me-donor27-story.html

Saving money by camping out on land adjacent
to the University of British Columbia

Fig. 13.5 Green commitment on another level - foregoing food and comfort to support what was then a leading environmental organization

Leadership means change and change will upset some people particularly those with a vested interest in business-as-usual. Some conflict is inevitable, so keeping the argument fact based is critical. Vested interests will attempt to polarize an issue to shut down useful discussion. What good are activist organizations or political parties which have stepped away from the issues they were created to address? What of the national conversation and democratic decision making?

Make sure the actions of the group you support are consistent with its original tenets and that they have a coherent platform mapping out a pathway to their end goal. To be credible and effective, activist groups and political parties need to have strong platforms on consumption and population with clearly defined steps to sustainability.

Role Models

Political leaders who have no burning interest in renewable energy or sustainability need to have examples placed in front of them to provide a starting point for their learning curve.

Compare the policies and action of those you are trying to convince to those of institutions which are world-leading such as participants in the UN's "Making Cities Resilient 2030" project representing 149 cities with 78 million residents.

Make all examples of excellent performance relatable and even better, use local illustrations. Are the communication links in place to enable effective improvements? What does MCR2030 suggest?

- Strengthening vertical links between local government with the national governments and national associations of local governments
- strengthening horizontal links amongst local partners to ensure sustainability
- connecting cities with cities to learn and share
  - Peer-to-peer learning has proved to be one of the most effective ways to achieve local resilience
- Find where your city is on the resilience roadmap. Start from there. Prestige works and so does shame.

Organizations with strategies spring up in many communities despite the failure of national governments to lead. One such initiative in Canada called ICECAP, is a six municipality organization in Central Ontario. It works through the 400 member Federation of Canadian Municipalities' Partners for Climate Protection (PCP) Program. The program consists of a five-step milestone framework that guides municipalities in their efforts to reduce greenhouse gas (GHG) emissions (Fig. 13.6):

The group has calculated the total "**Corporate**" emissions of the operations of each ICECAP member Council (e.g. administration buildings, public arenas, fleet of vehicles, etc.) and the "**Community**" emissions from each participating municipality. With data collected through their Carbon Calculator, ICECAP is the first in Canada to include recreational boating which is a significant contributor to local GHG emissions. Would a national model pick up that kind of detail? Unlikely, and it shows the importance of community based initiatives.

The path to emissions reduction is clearly laid out below (Fig. 13.7).

Fig. 13.6 Community climate action and planning is critical to drive the renewable energy transition

Fig. 13.7 A simple direct plan to monitor the process of de-carbonization

Going Broke on Growth: The Oil Sands Syndrome

The culture that has grown from five hundred years of biophysically decoupled growth has delivered huge benefits to humanity but is unsustainable as we now entering the depletion stage of most critical resources. Our current belief system has to change from boundless faith to one of analysis so we can begin to live from the world's sustainable flows as opposed to simply mining its finite stocks.

But today, most business and political leaders believe that size and rate of growth equal economic and social health. A close look reveals this not to be true. Typically, the largest and fastest growing cities have the highest taxes, the largest debt, the highest cost of living and the lowest quality of life. Growth pays off well for some but not for most people. As well, some business models simply don't work at all.

The oil sands is one of the best examples of a failed business model that exists solely because of its size and importance to a local economy. It is only profitable if one does not include its legacy costs in the equation. If the producers had to pay for the complete cleanup before they took out their profits, the project would have been abandoned decades ago.

The oil sands is the dirtiest and highest cost major source of oil on the planet and the most likely to be stranded by rising carbon standards. Despite this, governments continue to support the project with subsidies and lax environmental enforcement because its top line revenues are so large.

Physicists, biologists and ecologists and experts in essentially every field of real-world science have stated with growing alarm that further growth is a biophysical impossibility on this planet. Yet the great majority of commercial economists, politicians and media corporation employees continue to rail on about the necessity of growth and, in their minds, the fact that it can go on forever.

One commercial economist estimated that our population could keep growing for the next seven million years at 1% annually – in their terms, a very modest rate of growth. But starting with eight billion humans and growing at 1% annually, the number of humans will exceed the number of atoms in the universe within 20,000 years. Such is the power of exponential growth and the absurdity of endless growth on a finite planet, or universe.

Politicians came to grips with exponential growth of the Covid-19 virus very quickly. Why can't they grasp the identical process as it applies to other biophysical issues?

Commercial economists and media corporation employees who promote endless growth aren't connected to reality; they are connected to a paycheque.

Sustainability Versus Profit

This is a false dichotomy because success does not mean exploitation or inequality. A good business produces a worthwhile product or service and lifts all boats while making a profit for itself. If it isn't profitable, it likely won't survive or invest in and develop new technology or more efficient means of production.

Steady state economy does not mean companies don't turn profits or grow, it means that the society as a whole puts no more pressure on the environment than can be sustained.[8] Learning, technology and profits are not constrained within a sustainable framework. But unlike today, enterprises can't mine the planet, show a short term profit and leave bills behind that have to be paid by future generations.

In a steady state economy, profit from any source does not dictate public policy; rather public policy determines the environment in which profits can be produced. Business leaders may not put profit ahead of the health of the planet but in order to maintain the viability of their businesses AND adapt to sustainable practices, there will have to be standards which apply to all companies to level the playing field. Businesses who want to do the right thing need to be protected from unethical or desperate producers who will operate at the lowest standard to maximize profits.

Leadership: What Does It Look Like? (Fig. 13.8)

New Zealand had perhaps the most effective initial response to the pandemic. Here is an abbreviated version of New Zealand's **Prime Minister Jacinda Ardern's COVID-19 lockdown speech** of March 26, 2020.

"Good afternoon
   Like the rest of the world, we are facing the potential for devastating impacts from this virus. But, through decisive action, and through working together, do we have a small window to get ahead of it.

Fig. 13.8   Jacinda Ardern tops the democratic charts for effective leadership

---

[8] https://steadystate.org/if-its-profitable/

I also said we should all be prepared to move quickly. Now is the time to put our plans into action.

(We must) Act now, or risk the virus taking hold as it has elsewhere.

We currently have 102 cases. But so did Italy once. Now the virus has overwhelmed their health system and hundreds of people are dying every day.

The situation here is moving at pace, and so must we.

We have always said we would act early. If community transmission takes off in New Zealand the number of cases will double every five days. If that happens unchecked, our health system will be inundated, and tens of thousands New Zealanders will die.

There is no easy way to say that - but it is the reality we have seen overseas - and the possibility we must now face here. Together, we must stop that happening, and we can.

Right now we have a window of opportunity to break the chain of community transmission - to contain the virus - to stop it multiplying and to protect New Zealanders from the worst.

Now is the time to act.

Our low number of cases compared to the rest of the world gives us a chance, but does not mean we have escaped. I do not underestimate what I am asking New Zealanders to do. It's huge. And I know it will feel daunting. But I wanted to share with you the stark choice we face.

New medical modelling considered by the Cabinet today suggests that without the measures I have just announced up to tens of thousands of New Zealanders could die from COVID-19.

The worst case scenario is simply intolerable. It would represent the greatest loss of New Zealanders' lives in our country's history. I will not take that chance.

I would rather make this decision now, and save those lives, and be in lockdown for a shorter period, than delay, see New Zealanders lose loved ones and their contact with each other for an even longer period. I hope you are all with me on that.

Together we have an opportunity to contain the spread and prevent the worst.

I am in no doubt that the measures I have announced today will cause unprecedented economic and social disruption. But they are necessary."

True leadership makes choices simple, clear and decisive. How would this read if the climate crisis was to become more severe suddenly and humanity finally realized it was riding up on the bottom of an exponential curve of an accelerating problem just as it did with Covid-19?

Fictional: Prime Minister Jacinda Ardern's Climate Crisis Speech

"Good afternoon.

Like the rest of the world, we are facing the potential for devastating impacts from rapidly deteriorating climatic conditions. But, through decisive action, and through working together, we have a small window to get ahead of its worst consequences.

I said we should all be prepared to move quickly. Now is the time to put our plans into action.

We must act now, or risk extreme devastation to our citizens and our way of life.

We currently have only a few forest fires burning through the countryside while several floods are also underway on the west coast. In normal times we could have faced these difficulties and felt we could build back. But so did Miami, Brazil and the Netherlands once. Now climate extremes have overwhelmed their infrastructure and thousands of people are dying there every day.

The situation here is moving at pace, and so must we.

We have always said we would act early. If environmental conditions continue to decline ever more steeply in New Zealand our food, transport and energy systems will cease to function effectively, our health system will be inundated, and tens of thousands New Zealanders will die.

There is no easy way to say that - but it is the reality we have seen overseas - and the possibility we must now face here. Together, we must stop that happening, and we can.

Right now we have a window of opportunity to break the chain of cascading climate crises - to stop it multiplying and to protect New Zealanders from the worst.

Now is the time to act.

Our broadly based energy and food systems, compared to the rest of the world give us a chance, but does not mean we have escaped. I do not underestimate what I am asking New Zealanders to do. It's huge. And I know it will feel daunting. But I wanted to share with you the stark choice we face.

New biophysical modelling considered by the Cabinet today suggests that without the measures I have just announced tens of thousands of New Zealanders could die on a monthly basis.

The worst case scenario is simply intolerable. It would represent the greatest loss of New Zealanders' lives in our country's history. I will not take that chance.

I would rather make this decision now, and save those lives, and adopt emergency conservation and renewable energy infrastructure production for a period of decades, than delay, see New Zealanders lose loved ones and their way of life permanently. I hope you are all with me on that.

Together we have an opportunity to contain the crisis and prevent the worst.

I am in no doubt that the measures I have announced today will cause unprecedented economic and social disruption. But they are necessary."

Leadership has the courage to recognize danger and takes decisive steps to limit the damage and get ahead of the problem. It surrounds itself with experts and good data. It prepares the public for the change it is about to undergo and does not seek to minimize the problem or announce false or half- measures. Leadership also learns and is able to make decisions on the basis of incomplete knowledge. Those whose performance fell short of the New Zealand's PM now know the following doesn't work:

- Trying to walk a fine line or ride the dragon
- Depending on the market for leadership. Commercial infrastructure will be critical but the commercial market can't provide leadership on something that does not register either in its cost structure or in its sales forecasts.
- Working without goals, strategies or current data

Great Thinker (3)

Personal challenges and social conflict are nothing new as Rudyard Kipling's poem "If" of 1895 so eloquently reveals.

"If you can keep your head when all about you
Are losing theirs and blaming it on you,
If you can trust yourself when all men doubt you,
But make allowance for their doubting too;…………"

Beginning the journey towards a sustainable society isn't quite the high bar test of character that Rudyard Kipling might urge us to pass but it does require the ability to stand up at the front of the field. It is no longer a lonely place to be because there is now plenty of company. But naysayers and indolence are obstacles which one must be prepared to step around and over. Will humanity come together and recognize the immediacy of the biophysical threats of climate change and resource depletion in the same way we came to recognize the danger of Covid-19 spread?

Knowing When Not to Quit

Not understanding the challenge leads to false starts and pre-mature stops (Fig. 13.9).

Possibly Coronavirus SARS2 will be with us in perpetuity much the same as the threats of diphtheria, whooping cough, polio, measles, mumps, and chickenpox for which infants are now immunized. Covid-19 could 1 day disappear altogether but that may only present a growth opportunity for another

Fig. 13.9 George Bush - OK, the timing was off a little and, to be honest, we really never did accomplish or even understand the mission (https://www.cbsnews.com/news/mission-accomplished-banner-could-go-on-display-at-bush-library/)

virus. Certainly the challenges of maintaining sustainability and resilience will never go away in our ever changing world (Fig. 13.10).

Biophysical threats aren't a static enemy; they are a constant dynamic – always changing and always with us. Covid-19 positivity levels in California in mid-June, 2021 when the state re-opened were low at 1 in 125 people, but less than half of the population had been fully vaccinated. Still, these statistics provided strong reasons for hope against the original variant. However, the 5 times more transmissible Delta variant was well-positioned to do more damage. Delta drove a third wave but other variants have now replaced it worldwide. As you read this, is Omicron the last variant?

Is it time to declare "Mission Accomplished" or is it time to lay the foundation for a better response for the next pandemic?

California Re-opening party - CNN. Mission accomplished?

Fig. 13.10 California re-opening party was like celebrating the end of a forest fire with a bonfire

Learning Can Never Stop

Vaccines work by preparing the body for an attack by an outside organism in two ways. The first is by creating front line tools for this specific threat and the second is creating a familiarity with the virus. Thus, the body is ready to combat the virus by quickly ramping up production of defence mechanisms and employing them according to strategies which it knows has worked.

Several new biophysical threats have presented themselves to humanity over the past several decades in the form of climate change and resource depletion. We have been slow to engage these threats and currently have no tools or strategies to defeat them.

Present centeredness is a human condition which was illustrated by the fable of the grasshopper and the ant. As the reader may recall, the diligent ant survived but the large-living grasshopper did not. Has Covid-19 given us the foresight we need to avoid a well-deserved grasshopper-like fate?

Covid-19 is a threat which we have met and, if not defeated, at least countered sufficiently for semi-normal life to resume for most people. Certainly societies have not been driven to the edge of collapse by the current pandemic.

Beyond Covid-19, it is safe to say, most countries are now much better prepared for the next pandemic. Their health infrastructures have been built out with more domestic vaccine and medical supply manufacturing capacity. Just as importantly, the leadership class is much more clear on what works and what does not in preventing damage to their societies and citizens.

The public in general will be much more supportive of the early decisive measures that would be enacted by progressive governments now that they've lived through the long, drawn out and very painful consequences of delayed responses, half-measures and even denial.

Humanity will experience successive pandemic threats but as long as the memories of Covid-19 remain vivid, our future responses should be much more effective. But can we apply the Covid-19 lessons to the much larger threats and change our behaviour enough to avoid self-inflicted societal collapse? Have we learned that business-as-usual is not a strategy that can deal with biophysical reality and its challenges?

Our preparedness both on a personal and national level will substantially determine how successful we are in mitigating the clear and present dangers the failing health of our planet represents.

14

How to Enjoy a Sustainable Lifestyle with Both Fun and Security on the Path Towards a Green Future

Whose Hand Writes the Future?

Mother Nature is under no obligation to modify her rules to include humans among the species she chooses to support in her biosphere. Whether or not we will be included in this club is strictly dependent on whether we decide to deal ourselves in. The Japanese have a saying to describe the responsibilities one generation has to the next. How many of us at this moment can honestly claim to be handing down a better rice field?

Expectations play a large role in happiness and, if our expectations are realistic, most people can reasonably expect to be happy in the post-carbon era. But getting there will mean many will have to trim their expectations. Currently commercial advertising greatly influences wants and expectations and is in conflict with the global need to consume and produce less.

At some point humanity will arrive at Covid-19 herd immunity. Whether this comes through our own efforts or the simple progression of a virus through unprepared populations, with all that implies, is an open question. Similarly, the transition to renewable energy and sustainability will see humanity arrive at the end of the process with sustainable levels of consumption. How large a population and what level of per capita consumption they enjoy will depend on how well we've prepared for this process.

Will the transition be a chaotic one of denial, half-measures and reactive stops and starts? That would involve massive damage to individuals and the social infrastructure. It would be best if it was one of clear and coherent action that would preserve personal well-being and the best elements of our society.

J. E. Meyer, *The Post-Pandemic World*, https://doi.org/10.1007/978-3-030-91782-1_14

Left unchecked, the pandemic would probably have killed no more than 0.5–2% of the global population which equates to 1.8–7.2 million deaths in the United States, 3.5–14 million in Europe, 200,000–800,000 in Canada, and 7–28 million each in China and India.

We weren't willing to pay that kind of price to maintain business-as-usual but, if we did allow Covid-19 free reign with all of the collateral damage, would we really expect there to be minimal economic damage going forward? Based on our experience over the past 2 years, clearly an unrestrained pandemic would have done far more damage to society than "merely" killing tens of millions of people. Social and biophysical shocks are not surgically isolated events.

We need to mitigate and adapt to climate change and scarcity because, unchecked, each of these on its own will exact a much higher cost than anything an unrestrained Covid-19 pandemic would have been capable of. Are we going to be proactive and get ahead of the issues or are we going to be dragged through the process and emerge much diminished in an uncertain future?

What Did We Learn?

Covid-19 snapped the world out of its routine, gave many people an entirely new set of priorities, and opened the door to a new way of thinking. Health and family became a central focus and so did staying in touch with the community around us. Things we took for granted or had lost, once again became very important.

The importance of nature and space was catapulted past the rush of "the rat race" as priorities were re-shuffled almost overnight. Those with means stepped away from the glass-tower, monetized world towards a lifestyle of natural balance because it was simply the safest place to be during the pandemic.

Most importantly, we learned about the vulnerability of our way of life; that our political and economic systems are not geared to deliver stability in an increasingly dynamic world. We have now accepted that our society has vulnerabilities. Some of them are social, where people will only respond so well to clear community needs and some are administrative where public policy apparatus finds it difficult to implement efficient and consistent programs.

These two things combine to underline why it is best to prevent problems and if that is not possible, to quickly get out ahead of them. Trying to walk a fine line of control with the very cumbersome and imprecise administrative and civil control mechanisms currently in place can lead to a series of damaging missteps.

Table 14.1 Strong Covid-19 measures were the best option for good social and economic outcomes

Early strong measures	GDP change 2019–2020	Covid-19 deaths
Australia	-2.5%	910
Taiwan	4.4%	550
China	3.0%	4600
New Zealand	-2.9%	26
	Average = 0.5%	Total = 6086
Denial, no or late measures		
Brazil	-4.1%	520,000
UK	-9.9%	130,000
USA	-2.3%	610,000
	Average = -5.4%	Total = 1,260,000

Hopefully, we also learned to pay better attention to biophysical threats as climate records fall in increasing numbers and extreme weather events are almost omnipresent in world news. Humanity may still be short of the awareness level we must have to drive toward the threshold of sustainability but at least we are beginning to look over our collective shoulder.

The table below compares the economic and mortality outcomes of two groups of countries. The first group recognized the scale of the Covid-19 threat and acted early and decisively. Countries in the second group initially denied there was a threat and they reacted by instituting a patchwork of half-measures. Of the three, only the UK finally came around to implementing effective measures but these came 130,000 deaths late. In addition the oblivious countries also ended up paying a far higher economic price (Table 14.1).

Business-as-usual simply wasn't the best path to follow when the pandemic struck. It will fail on an even grander scale when the consequences of environmental decline and global warming arrive on our doorstep in force.

Surely beyond learning how to make vaccines quickly, there is a larger systemic lesson to be learned from our Covid-19 experience that involves realigning public priorities to meet clear threats.

The Challenge

In comparison to achieving sustainability, Bill Gates, for one, sees dealing with the pandemic as "very, very easy".[1] Further, we should not underestimate the scale of the challenge because as Gates notes, "We've never made a

---

[1] https://www.bbc.com/news/science-environment-56042029

transition like we're talking about doing in the next 30 years. There is no precedent for this."

Perhaps the World War 2 mobilization suggests we are capable of reacting strongly once given the motivation but when will we have the motivation? What has to happen to get us to sign up for front-line climate change mitigation duty?

Our approach has to be multi-faceted and cover everything from electrification, to efficiency, to conservation and recycling on the human activity side to rebuilding forest heath and soil quality with initiatives like biochar.[2,3]

No society is stronger than its soil and restoring its health has to be one of many high priorities. Erik J. Joner of Norwegian Institute of Bioeconomy Research[4] cites the advantages of adding charcoal back into the soil as follows: "In general, agricultural soils supporting cash crops benefit most from BC additions through improved water holding capacity, improved tilth, reduced phosphorus fixation (making added P more plant available) and improved water infiltration during heavy rain."

Biochar enrichment isn't suitable for all soils. As Alford-Purvis points out, simple compost might be better in many cases and and biochar needs to be applied as a measure tuned to specific circumstances through careful research and management.

Biochar, like the Clean Cities Initiative[5] is a comparatively narrow issue but it represents the scores of initiatives which must be undertaken to re-establish our society on firm biophysical ground. When you watch the videos of homes floating down rivers during floods, keep in mind that the rivers are brown because they are filled with soil.

Most of all on the biophysical side, we need to change our reckless behaviour enabled by the seemingly unlimited possibilities presented by cheap fossil fuel. On the decision making side, we need to institute real world metrics to replace the financial system which has severed any connection between the biophysical world and national policy formation.

To sum up we need to:

- Change our habits
- Reduce our consumption
- Change primary energy systems

---

[2] https://www.nature.com/articles/517258a?proof=t
[3] https://www.biochar-industry.com/why/
[4] https://nibio.no/en
[5] https://www.motorweek.org/features/auto_world/green-reservation/

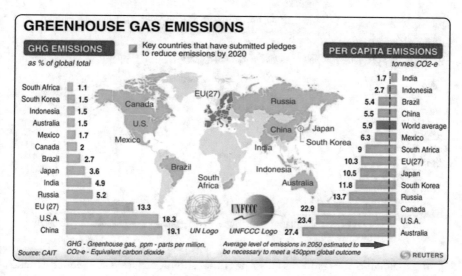

Fig. 14.1   GHG emissions by country and their per capita emissions targets present a very large gap to be crossed. (Source: CAIT)

- Clean up past mistakes
- Rebuild our forests
- Rebuild our soil
- Radically enlarge natural regeneration zones
- Adopt biophysically based national goals and metrics
- And be well on the way to implementing it all in 30 years.

Yes, that is a challenge.

The graphic below shows where we are and how far we have to go (Fig. 14.1).

Clearly we have to cover a great deal of distance and the sooner all of us engage and get going, the better the outcome we'll be able to achieve. If we don't do it ourselves, Mother Nature will dip her hand into the raffle bowl of outcomes and hand us our collective fate.

Constraints

Change will bring new challenges and the increasing scarcity of many key raw materials needed for conversion of our society from fossil fuel to electricity will impose constraints. But these are constraints, not fatal flaws, unlike fossil fuels whose negatives of lethal climate change and the visible end of useable reserves make their long term use impossible.

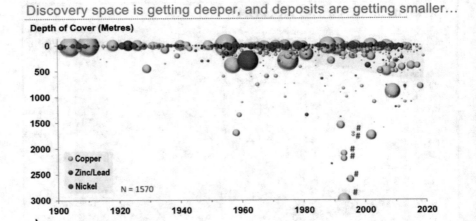

Fig. 14.2 Depletion of resources illustrates dis-economies of scale at a glance. Smaller, harder to reach deposits

As previously discussed, ore deposits are declining in richness and we now have to dig deeper for smaller deposits than ever before. The chart below illustrates that the deposits of copper, zinc and nickel, so critical for electrification, now lie deeper below the surface and are smaller than during the glory days of development in the last century. On the left side of the chart we see larger deposits near the surface in 1990 and in just 30 years the trend is one of smaller deposits buried deeper (Fig. 14.2).

Further, electrification and renewable energy require larger quantities of these vital minerals than we have ever extracted before. Recycling, assumed in the exercise that produced the graph below to be 55% efficient, is critical and rates need to exceed the 90% mark to maintain a moderately reasonable material and energetic standard of living. Mining space is an energetic non-starter (Fig. 14.3).

The above graph shows that converting to all renewable energy will require an increase in copper production from 1.1 million tons annually to 3.2 million tons. This does not include the copper needed to produce electric cars and motors of all description, nor does it include the copper necessary to build out the electrical grid. However, increasing the capacity of the grid may not require a huge amount of additional material if solar and wind systems are well-distributed.

But the chemical/electric batteries which will store energy in countless applications require production to be 12 times higher for aluminum, cobalt, lithium and nickel to limit global warming to the 2 degree C mark. Missing

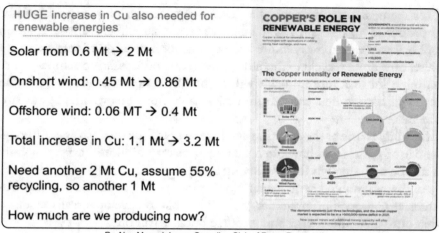

Dr. Alan Menzel-Jones, Canadian Club of Rome Presentation

Fig. 14.3 Despite the dis-economies of scale, we still need many Mt (Megatons) of minerals to build renewable energy systems

that and sliding to a 4 degree increase will require just double the current production.

This is not to imply that a 4 °C warming scenario is relevant since no one has suggested that human civilization at any scale is possible if the earth's atmosphere warms another 4 degrees. It just makes clear that failing requires less effort than succeeding.

> **Constraints Versus Brick Walls**
>
> Seasonal storage of energy in northern climates is simply not possible with any electrical battery technology on the horizon due to the immense amount of materials required.
>
> Geothermal storage is the only option. Hydrogen might appear to be viable but it has distinct disadvantages. Also, it can only be made using high grade energy like electricity which will likely be scarce.
>
> Geothermal uses low grade energy – heat – which is much more abundant and easier to harvest.

Humans have already picked the lowest hanging fruit and it is taking more work to find viable deposits and longer to get them into production. Higher environmental standards are needed to minimize cleanup costs. Adequate standards will reduce the large financial burdens being imposed on taxpayers for mine cleanup long after the ore and the company that mined it have disappeared.

If a material or product can't be recycled then, in the long term, it shouldn't be produced and herein lies a huge problem that offers no easy solution. Sophisticated electronics require a large range of rare earth minerals, some of which cannot be recycled. Therefore, any new supplies must be mined. Since mining a lot of a rare minerals will eventually result in effective depletion, there will have to be a constant re-design of components which use available minerals.

Mobility is one of the great life enhancements of the modern age. Although the size of vehicle and how often we take trips may diminish, expect mobility to still be high enough for a satisfying lifestyle.

As the renewable energy generation infrastructure is built out along with extensive conservation upgrades, we will be able to develop a realistic idea of what our per capita energy and resource budgets can be.

Long Term Trends

Below is a listing of what would seem to be fairly safe predictions from the perspective of early 2022.

If an industry is dependent on growth to sustain it, then it comes with an expiry date over a time frame of several decades. If growth dependent sectors are still going strong beyond that, perhaps human society itself should nail up a "Going Out of Business" sign.

- What is certain to happen:

  - Population stabilization and decline will be occurring in most of the world over the next 15 years. Regions with continued population growth will likely be sinking into social turmoil.
  - Extreme weather will place increased pressure on the viability of many well-established communities. No regions are immune from this trend but some are more exposed than others to floods, drought and rising ocean levels.
  - Commodities will become more expensive
  - Some governments will not attribute cost increases and shortages to their real biophysical causes. Instead, they will point the finger of blame at political rivals and other nations rather than implementing science based programs. Their responses will be ineffective and counterproductive.

Once we are into the post consumerism and growth era and sustainable levels are achieved, what will the landscape look like?

- A greatly depleted but cleaner and recovering planet
- Higher domestic energy system costs
- Less disposable energy
- More secure and reliable energy systems
- More expensive materials
- Lower material consumption

  - longer life products
  - higher quality
  - upgradable
  - more easily repairable

- Less mobility and less need for mobility
  - far less air travel
- Higher recycling standards
- Less foreign foods
- Less foreign goods
- Broader based economies
- Full employment
- More egalitarian societies

If we have managed to learn something from past commodity crunches, our coming resource crunch won't involve shooting at each other or invading other countries. Large-scale war throws all of the cards up in the air but it may now be a less attractive option for desperate governments given the destructive power of current weapon systems.

The path towards sustainability will be more difficult in the near term. But if we fail to work towards a balance with nature, the path we are on will become very difficult, very quickly at some, not too distant point in the future.

Transitions

The transition to sustainability will involve changes in the way we think, measure and live. Thinking will shift away from consuming the maximum amount of products to making the best use of the products to do the job that needs to be done. We will measure the entire lifecycle of our activities.

We'll be more aware of our surroundings and less self-centred as our attitudes change one by one. These changes range in scale from the very large to the very small aspects of life:

- Realizing that we really do want to become a good ancestor
- Using half the toothpaste you used to
- Personal vehicles are idle 93% of the time which implies a potential huge downsizing of the personal transportation fleet. Where will autonomous vehicles come into the equation?
  - Everyone needs a truck from time to time but very few people need one all of the time. Is it time for a Truber service to enter the market?
- We will NEED-size not WANT-size in our decisions.
- A fast charging network and efficient EVs with 20 kWh batteries could move 8 times the number of people that one single Hummer with a 200 kWh battery typically could.
- If we can't recycle it, we don't make it
- Shift in taxes from an income to a consumption based system
- Shift in tariff walls from zero in the free-trade globalism era to very high import duties of 100–200% on all imports. This will increase domestic production, broaden national economic bases, reduce emissions and support consumption based taxation.
- Learning to get out of the way and simply let Mother Nature do the heavy lifting at her own pace.

We need to embrace naked food by shedding the single use packaging which constitutes a significant source of emissions and pollution. But these steps are not easy:

- BYOC - bring your own container
- Don't buy anything that comes in disposable packaging. No containers, bags, wrappers or jugs.
- One-stop shopping goes out the window and is replaced by visiting specialty shops which requires more planning and time. **Ditto buying in bulk, farmers markets, local bakeries, delis and specialty shops. If it travels it has to be packaged!**
- **Fruits and veggies are easy as no packaging is required. Use your own bags until you get home.**
- **Anything instant or frozen needs packaging so fresh is always better.**
- **Cook more at home as** it tastes better and is cheaper and healthier.

Standards will change. We have standards for many things from civil conduct to speed limits to building codes and emissions. People know that to violate these means they are off-loading the cost of their recklessness onto other individuals. Everyone who understands the reason why you are not allowed to burn rubber tires in your backyard can grasp the necessity of both wearing masks during a pandemic and converting to clean energy to avoid a climate change induced societal meltdown.

Science won't scare many people anymore because we will trust our sources. "That is the way science works. You work with the data you have at the time. It is essential as a scientist that you evolve your opinion and your recommendation based on the data as it evolves. It is a self-correcting process." - Anthony Fauci. We have to be prepared to be shown that we are wrong or only partially right and we always have to be prepared to learn.

New World for an Evolving Humanity

Evolving and adapting means taking the demographic, energy and sustainability transitions in stride to avoid solving domestic tensions with either domestic or foreign wars. We need to adopt cooperative development with our neighbours and live within our means and borders. This will allow us to stop fighting each other as well as the planet.

But some habits are going to be very hard to quit. Cars are a personal joy but also an addiction and an environmental disaster. Mobility and convenience are highly coveted luxuries and, the way our cities have evolved around cheap and fast mobility, often necessities.

Cities are working to reduce car traffic but unfortunately, the pandemic has made mass transit less safe and less popular. Smaller cars can mean more lanes of traffic and smaller parking lots. This all translates to better flow. But do the wealthy want to rub shoulders or fenders with the masses? Will people who shun first class air travel in favour of private jets tolerate driving in something that doesn't weigh twice as much as any of the vehicles around them?

Maybe "will we have adapted" isn't the real question. Maybe the question is "how adaptable are we?" Our hubris has reduced our adaptability but the pandemic has taken our pride down a few pegs. As biophysical threats loom ever greater, if we keep our heads, we are likely to become much more adaptable.

Our highly stratified society is now dominated by elites who are totally dependent on population and consumption growth. Whereas leaders of a tribe could lead change for the greater good easily because it was also in their personal interest, current leadership will be effectively cutting their own

Turnkey Decarbonization Package Deals

Fig. 14.4 The commercial market may step up to make decarbonization a logistical breeze for some individuals

throats to serve the public good. Sustainable practices will offer enormous commercial opportunities but just not to the current elites (Fig. 14.4).

Developers are beginning to offer turnkey no-carbon homes and neighbourhoods. How long will it take for them to offer complete package deals in which the customer steps from a large vehicle, large house, fossil-fueled lifestyle to one that is carbon neutral and lower consumption?

Pack your suitcase, dump your stuff and walk into a completely outfitted lifestyle rather than going through dozens of decision making processes to whittle down your footprint over several decades. Convenience and clarity can be developed in fairly short order as the Whisper Valley development clearly demonstrates. How quickly will this trend develop?

One of the key adaptations will be the effective fusing of government mandated priorities and time horizons with the commercial market system. Once standards, as opposed to short-lived incentives, are enacted which will demonstrate clear and stable cost trends, consumers and companies can make their own energy and financial rational investment decisions with confidence. Companies can invest in business plans and equipment which will allow them to offer new products and services which will make it much easier for the consumer to say" Yes!" to downsizing and decarbonization (Fig. 14.5).

Fig. 14.5 Do you need to dominate your environment? Why?

Will we dominate nature or live in it? Humans have fought their way to the top of the food chain but surely spiking the ball and doing cartwheels isn't necessary at this point.

Micro EVs, with their 100+ km ranges, are all most people need in order to do what they need to do. As we move out of the global assault mode to become good ancestors we can please just stop re-conquering the planet and let it get off the mat?

People, who will not look around them and only change unless they are offered something which is better, cheaper and more convenient, may become befuddled by changing events and standards. If you can't help them, step around them (Fig. 14.6).

Micro EVs

IMA
Colibri
110km

Artega
Karo
200km

Tazzari
Zero
200km

Microlina
200km

Fig. 14.6   Micro-EVs are all most people need. And they are easier to park

Will We Adapt?

The signs have all been there for many years but so have the roadblocks (Fig. 14.7).

This banner in California presents a simple and clear message.[6] If we want to have supplies last, we need to learn to live on a budget. For several centuries, fossil fuels made resource budgets irrelevant, but with their decline, the need to think about the larger picture will return. Has the pandemic restored in us the awareness we need to once again take active control of our destiny?

"Waste not, want not" may become words to live by once again and wretched excess may come to be seen as a very distasteful public exhibition by most. But still, some people enjoy making ostentatious displays and other people make their living selling it to them.

Business as usual will be going out of business but not without a fight. With aging and localism trends in place, productivity and wages are likely to increase while housing inflation will decline producing a higher level of equality. Socially, fiscally and for most individuals, these are positive trends.

But not everyone will be enthralled. The value of real estate in the United States is approximately $50 trillion. An annual 5% rise puts $2.5 trillion in valuation into play. Most of this is in the hands of individual owners but hundreds of billions accrue to the balance sheets of speculators. No matter who

Fig. 14.7 Use less water. Use less energy. Use less material goods

---

[6] https://www.cbsnews.com/video/record-heat-wave-hits-western-states-gulf-coast-prepares-for-possible-tropical-storm/#x

owns the properties, banks stand to be involved at some point by providing ever-larger mortgages.

The figures for Australia and Canada are $6 trillion and $4.8 trillion respectively. The interests which play in the finance and real estate sectors dwarf the players in the armaments, tobacco and oil sectors. Finance and real estate are absolutely dependent on the inflation of property valuations.

From 1990 to 2020 the market value of residential real estate increased by 5 times (that is 500%), in Australia, 4 times in Canada, 4.1 times in the UK and 2.9 times in the USA. There will be extreme resistance to shutting down the conveyor belt of money from real estate inflation to the pockets of the rich.

Housing inflation is essentially a tax on younger generations via a wealth transfer from the young to the old and to speculators.

In growth focused economies, efforts to change the belief that a house is an investment should always increase, to the recognition that a house as a long term consumer good, will meet fanatical resistance in the corporate media.

Despite the spin and intimidation being employed to maintain business-as-usual, public awareness of the declining state of our planet and its knock-on effects on their daily lives is now high and will continue to grow. At some point the demand for effective action will become impossible to resist (Fig. 14.8).

The California wildfire trend is part of a world-wide phenomenon. In some areas the average area burned is declining but in others it is increasing. At the moment, we seem to be able to re-build and recover from natural disasters for the most part but, at some point, if these trends continue, the social fabric will

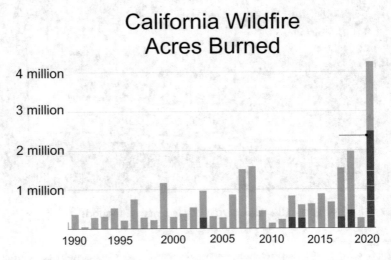

Fig. 14.8 California Wildfire Acreage. At what point do forests turn to grasslands?

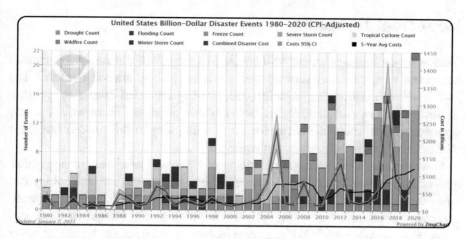

Fig. 14.9   Extreme weather costs are rapidly climbing

not be able to repair itself. As extreme events become more frequent (5 of the biggest 10 fires in California happened in 2020) recovery has to give way to mitigation as our primary response (Fig. 14.9).

The costs of floods, fires and droughts[7] are adding up in dollars but they are also taking a toll on public confidence. Unlike with the pandemic, decades of strong climate mitigation efforts may only prevent the situation from getting worse decades later, providing little in the way of positive reinforcement for restriction weary citizens.

The near constant stream of extreme weather events in 2021, with record floods, fires and droughts occurring simultaneously was enough to regularly elbow the pandemic off the top headline spot. There is a very visible climate crisis and there is also a less visible resource crisis. Can humans and their political systems recognize these threats and put institute programs to deal with them on the front burner? Can we adapt and overcome?

The IPCC (International Panel on Climate Change) report of July 2021 is unequivocal in its warning about continued complacency. In this 8 minute video,[8] the highlights are spelled out. Would you ignore a message like this from your doctor?

---

[7] https://www.ncei.noaa.gov/news/calculating-cost-weather-and-climate-disasters
[8] https://www.youtube.com/watch?v=1J0lCBjMgvg

New Deal Template

Which forms of government will be able to negotiate the transition from a consumer to a caretaker society over a relatively short several decades? The task involves maintaining equality, minimizing stranded assets, supporting declining sectors and making life better where possible. There will be many opportunities to make life better for the vast majority in this process but there will be sacrifices from everyone at some point.

Governments demanding unrelenting sacrifice will need very strong police and military cohesion to stay in power. The assumption that technology can smooth out most of the bumps along the way may instill hope but a well-grounded approach is best. As Oliver Cromwell put it **"Trust in God and keep your powder dry"**. We can trust in human creativeness but must never lose sight of the world around us.

Regardless of the form of government, large changes have to take place and the disruptiveness of these changes will be closely related to the time frame in which they are implemented. We are a good 30 years late in starting to reduce carbon emissions. The graph[9] below presents some of the trade-offs between the magnitude of the changes we need to implement to hit different warming ceilings and the time frame required for different mitigation schemes to have the needed effect (Fig. 14.10).

The above graph illustrates the theory, not the robustly modelled fact, of the performance of different levels of carbon reduction strategies. In order to

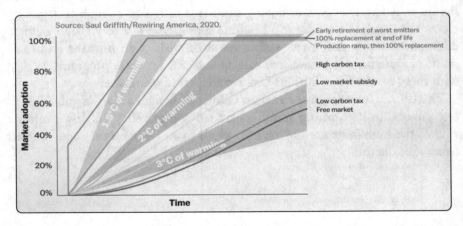

Fig. 14.10 Slow starts and delays cost more and make outcomes less certain. (Source: Saul Griffith/Rewiring America, 2020)

---

[9] Decarbonization USA, https://www.youtube.com/watch?v=QfAXbGInwno

hit the 1.5 °C target, we will have to take drastic immediate action. 2 °C requires very strong direct measures in addition to very high carbon taxes. 3 °C is the result of a laisse-faire approach and 3 °C is not necessarily a level of warming that will support a moderately sophisticated human society of any scale. How quickly and efficiently this can unfold will only become clear once we are well into it.

If we had started 30 years ago with carbon taxes and incentives, we would be well along the way to achieving a net zero carbon emissions without the need for drastic interventionist measures. But now, having pumped far more carbon into the atmosphere in the past 3 decades than we needed to, our remaining carbon budget is very small and our time remaining is very short.

These conditions make it certain that governments will have to enact drastic measures as they did with the Covid-19 response to avoid a far greater disaster. Market mechanisms are now far too little and decades late too late to be effective on their own.

The Timing-is-Everything graph above, although based on as much hard data available is an educated estimate of the time frames by Saul Griffith who describes it as "more phenomenological". In his book, "Electrify - An Optimist's Playbook for Our Clean Energy Future "he offers what he feels to be a realistic and feasible action plan for fighting climate change while creating new jobs and a healthier environment. Once again, the need to create detailed mathematical modeling of real assets and processes is clear.

The automotive industry can serve as an example as a sector which will be subject to rapid change. It will be an asset teeter-totter of sorts when ICE manufacturers go down and EV manufacturers go up. There is more involved than market share loss. Huge amounts of capital in the form of tooling, parts, repair facilities, test facilities and intellectual property including personnel will be stranded and eventually written off.

Legacy manufacturers will find themselves with productive capacity with nowhere to sell the output as EVs approach the price level of gas models and then perhaps undercut them. In fact, given the performance improvements and lower operating costs, price parity is not required for EVs to displace ICE vehicles. In Norway EV sales are well over 50% and in Germany and the UK, the figures are 25% and 12% respectively. In the UK, EV sales in 2020 increased 189% over 2019. This was from a starting point of near zero in 2015.

According to analysts at Morgan Stanley "the value of the businesses related to ICE will be worth near zero, if not negative".[10] This is a similar situation to that of coal mines and plants, the oil sands and numerous other industries

---

[10] https://thedriven.io/2021/06/10/petrol-car-makers-may-soon-be-worth-nothing-or-worse/

which are dependent either on fossil fuels or never-ending population and consumption growth. Whether electric or fossil fuel based, there are simply too many cars and houses and people using too many resources. This will change.

In the case of private enterprises, it is not enough to be right, it is necessary to be right at the right time in order to survive and prosper. Governments have to be right but as long as they are on the right path, timing is far less of an issue.

Which Governance Models Survive?

We've built our societal house on the sand of fossil fuels and we will have to re-build it on the bedrock of renewable energy. The economic and political power structures which have taken root with the explosive growth fossil fuels have enabled will undergo change as great as that of the energy sector during the transformation to a sustainable society.

Any governance model can survive if it provides the great majority of its people with security, fulfillment, health and realistic hope. The common ingredients needed to achieve this are investing for the long term, high levels of equality and a very firm grip on the biophysical world within which the society operates. Most governance models can succeed and all models can fail.

However, throughout history, the common pattern for governments and rulers is that of increasingly stratified societies whose elites slowly become disconnected and cease to talk to their constituents or acknowledge their interests. Inevitably, the seeds of discontent are sown and once a climate or resource disruption occurs, the ineffective leaders and their courtiers are replaced.

Economic resilience stems from a broad economic base, reliable supplies of food, energy and vital goods and services. Social resilience requires a transparent national conversation which involves most major groups and prepares the public for change. This prevents shocks from ringing through the social structure and causing a loss of faith in the system.

Authoritarian regimes may stimulate a national conversation but this is likely to be mostly one-way. As long as the life experience of most of the population is positive, little national conversation may be needed and rigid enforcement measures can be used in place of public forums. In democratic societies, a high level national conversation is needed in order to keep the majority of the population onboard with government initiatives.

Both the above statements assume the governments are doing their best to assure the welfare of their populations. Malfeasance, given enough time, will bring down any form of government no matter how brutal the media and military compliance measures might be.

How far ahead in generations is it necessary to look to establish a reasonably good chance of enduring social stability? The Iroquois based their decision making on a 7 generation horizon but they were already operating within natural boundaries. Perhaps our modern societies need to base their planning horizon on the goals of achieving sustainability and only use time to define the steps necessary to reach it. Whatever the timeline needs to be, it has to be measured using a biophysical clock and not the stopwatch of the commercial economy calibrated with the quarterly bonuses of executives and the election cycle of politicians.

Prepped but Not Isolated

Preppers are people preparing for the disintegration of our sophisticated societies. They assume there will be social upheaval and shortages of critical goods and commodities following a large crisis. Their concerns go beyond toilet paper. John Ramsey, who heads an online community of preppers in North America called The Prepared,[11] covers a number of topics on his site including:

- How to make, store, and eat hardtack.
- What common or household objects will stop bullets?
- Beginner's guide to gold, silver, and precious metals.
- Bug in vs. bug out: Why your home is always the default choice.
- Best foods to grow in a survival garden.
- The 10 top crops when you need to grow your own food to survive.

He attributes the rise in interest in his site to the increasing number of natural disasters due to climate and the seeming inability of US government and institutions to manage their consequences. While the current Covid pandemic is not "the big 'un" in terms of disasters, according to Ramsey, it has highlighted the weaknesses in how we live.

But sustainability cannot survive in isolation, it must be broad based. Preppers may initially be able to do better than most people after a major shock but, long term, their fates won't be substantially better once coherent government and the ability to maintain essential social services are lost.

---

[11] https://theprepared.com/

People who wish to invest their time in building up their own level of sustainability along with that of their community, can make great strides without re-inventing the wheel of survival. Those who have invested in hard productive assets and the health of their community are likely better equipped to live life and remain secure than those living in isolation and preparing for societal collapse. Families with an energy positive, low demand house, efficient vehicle and living in close proximity to a resilient food producing area with a cohesive community are as prepped for the long haul as any survivalist can ever be.

A real community features trusted sources of information and trusted leadership which have the same goals and aspirations as the general public. Trust is a core concern for preppers and not an unfounded one.

No expert will claim it will be easy but the replacement of fossil fuels is inevitable as the following comments from biophysical economics guru Charles Hall, makes clear.

"In a comparison between an oil well and a wind generator at year 1, the oil well produces many times the output of a wind generator. If it is a fracked well, its output drops by almost 70% annually and it is only a few years until the wind generator out produces the fracked oil well.

In terms of a conventional oil well, we almost aren't drilling those anymore because conventional reserves have almost all been tapped and substantially drawn down. The oil well that produces over 60 years and then starts to decline at a rate of 5% annually is well in our rear-view mirror. If these were available, we wouldn't be fracking or digging up the oil sands.

At this point, peak oil had occurred on 6 of 8 continents and some 38 of 46 oil producing countries (Hallock et al 2014). Conventional oil peaked over a decade ago and now we have been reduced to boiling our oil out of sand and breaking up deep rock formations and sucking the oil out through the cracks. Yes, I'd say the best days of cheap oil are well past us."

There will be some clouds and light showers drifting across the bright blue sky of the renewable energy future. Once fossil fuels have been almost completely replaced, incentives for electric cars and devices will fall away and the tax burden will have to be borne by the users of all manner of electrical devices which use the roads.In the case of EVs, since charging will take place mostly at home, fuel taxes simply won't pay for road network upkeep. Governments will be faced with either charging an annual license fee, probably in the range of $1000 or charging on a kilometer basis, if the populace is comfortable with the monitoring this would require.

Beyond tax structure shifts, there is a host of questions for which we currently have no answers such as sourcing all of the new materials we need to run our electric world.[12]

On a much larger scale, the successful transition to renewable energy will not diminish the other major biophysical threats from biodiversity loss to soil degradation. Neither will it stop the buildup of carbon we've put into the atmosphere from continuing to increase temperatures for decades. But still, decarbonization of our society represents a critical turning point both biophysically and in our relationship with the planet.

Once We've Arrived

One can revel at least a little bit having put one foot outside of the rat race and knowing the entire nation has turned a page towards harmony with the earth. Prosperity can continue indefinitely by focusing on the endless potential of developing humans rather than exploiting finite nature.

Covid-19 has demonstrated the real cost of attempting to maintain business-as-usual instead of dealing directly with threats. The larger issues of climate change and resource depletion have to be seen, not as challenges to be defeated, but as inevitable transitions which must be managed. It is up to us to determine how well we will be faring on the other side of these processes.

> Our #1 duty is to prevent the most stable period of climate humanity has ever experienced, and upon which our civilization depends, from coming to an end by our own hand.

The Covid-19 pandemic has exposed many to a new perspective. The world now looks different through the lens of a health crisis so it will through the lens of the energy transition and sustainability. As an energy thinker, you can recall your past habits and those of your family and neighbours with some level of bemusement. Our culture has committed many sins of recklessness and wretched excess but these are sins of societal adolescence and will fall away as we mature into a sustainable society.As an individual who has taken the first large steps towards a post carbon world you will have electrified and downsized which will have reduced your carbon footprint by 70% to 80%.

---

[12] https://www.scientificamerican.com/article/new-wind-turbine-blades-could-be-recycled-instead-of-landfilled/

Once your own energy security is assured and costs are capped, the machinations of world events will be much less of a threat. Hopefully you will be in a community of equally healthy, productive and secure people who recognized early that investing for their own future meant doing the right thing for the planet.

A society that has been able to learn to adapt to nature, rather than attempting to dictate to it, is one that will endure. The quality of life of its citizens will be enhanced by the increased confidence energy and real cost stability brings. The satisfaction that comes from passing on a tradition of learning to forge a closer connection with the natural world to the next generation can't be expressed in dollars.

Appendix

Interesting Innovations and Technologies with Promise

Some Rough Notes

Humans are always looking for new frontiers as well as the path of least resistance, not to mention a quick buck. Over the past several centuries, technological breakthroughs, subsidized by the ever greater availability of fossil fuel energy, have provided much easier ways forward.

As biophysical challenges mount, we may be looking less for ways forward than for ways out of our problems. Will the innovation cavalry ride over the hill just in time to save us from disaster or perhaps even significant inconvenience?

Will we be able to get back to consuming more and more without the least concern for downstream impacts? That is highly improbable.

Any new technology requires a full lifecycle evaluation. We've seen the outcomes from projects which have appeared to be saviours because of their size. Ethanol, the oil sands, fracking (??) asset inflation etc., all represent huge cash flows. But do they make sense financially, energetically and environmentally once all of the bills are added up?

Some things once made sense but possibly do no more. Suburbs and urban sprawl were enabled by cheap fuel and a surplus of farmland. Will these conditions that created sprawl continue to exist going forward?

© The Author(s), under exclusive license to Springer Nature Switzerland AG 2022
J. E. Meyer, *The Post-Pandemic World*, https://doi.org/10.1007/978-3-030-91782-1

There may be massively more advanced energy systems about whose existence we currently are completely unaware. Maybe those flying chicklets are real and have both mass and volume. If so, they are powered by energy systems beyond anything our current level of science has even hinted could possibly exist.

Above all, interesting doesn't mean practical or even functional, now or at any time in the future. These are options to ponder and certainty not to blindly bet on. There are many "new "technologies which are really nothing more than schemes to attract investor dollars. Once that is done, they evaporate into techno jargon and then disappear altogether.

But clearly, new and worthwhile developments will come along which will do even more than their early backers claim. In any case, worthwhile or not, many new ideas are interesting and worthy of exploring even if they do nothing than stimulate interest or provide a new way of looking at an old problem.

When does "interesting" technology become "useful" technology?

- EROI higher than current technologies
- Low maintenance
- Broadly accessible
- Highly recyclable
- Low resource consumption
- Compliments other energy generation and storage systems
- Solves clear problems

For energy generation, higher EROIs define the threshold of worthiness. Massive, complex and hard to maintain systems may generate some energy but do they generate more than that used to create, maintain and recycle them?

That is the hurdle every new technology and design must pass. How will the developments listed below pan out over time? It is impossible to tell but, at this point, they seem to be addressing a problem that would be worth solving.

Interesting Technology Categories

Mobility

- Cars

Structural Batteries Batteries weigh a lot and if they can be incorporated into the support structure of vehicles, it will allow the weight of the chassis to

be   reduced. https://www.reuters.com/business/autos-transportation/
understanding-structural-ev-batteries-2021-07-23/

Souped Up Electric Vehicles Will no emissions restrictions push the number of electric tuners past that of ICE tuners?? Getting past emission requirements for internal combustion engines is a large impediment to every backyard
mechanic and startup with an idea on how to make transport better. Electric
drive systems remove that obstacle and hopefully open up the field to a wide
range of innovations and small, custom manufacturers. And racers will be able
to apply their learning on the track to mainstream products once again.

https://www.youtube.com/watch?v=s7n9CPflu_c

Weird and Wonderful Electric Options

Portable Electric pizza option to plug into V2G electric motorcycle. Forget
the charcoal, starter fluid and barbecue, just plug in your lightweight pizza
and burger oven for roadside lunches.
    Electric Crate motors are now available to electrify your old V-8 classic
car – silently. Ford and Chevy e-Crate motors fit anywhere a V-8 would.

– https://driving.ca/auto-news/technology-news/
    crate-news-ford-teases-electric-crate-motor-so-you-can-ev-swap-
    your-classic

• Boats

Silent Yachts

Given their large surface areas and the desire for quiet and natural experiences
on the water, yachts would seem to be a logical application of solar and electric drive technology. Simply pasting solar panels onto existing designs though,
has proved awkward. But the application makes a lot of sense and here is one
company who designed a yacht to be solar driven from the outset.
https://www.youtube.com/watch?v=JHqwPH50AFo

• Space Tourism
    This is a predictable application of advanced technology but it involves
burning off the same amount of carbon as a two way trip from Singapore to

London in only 8 min. It is easy to understand the thrill but do we really have to do this?

- Building
  Smart Home Surfaces - Light and Heat control.
  White paint that reflects 98% of sun's energy

— https://www.bbc.com/news/science-environment-56749105
  Transparent wood: Researchers at the University of Maryland have turned ordinary sheets of wood into transparent material that is nearly as clear as glass, but stronger and with better insulating properties. It could become an energy efficient building material in the future.

— https://www.cbc.ca/radio/quirks/
  scientists-develop-transparent-wood-that-is-stronger-and-lighter-than-glass-1.5902739

Smart Home Energy System Controller

Span, a San Francisco-based company has begun commercializing the smart circuit panel. This creates a single point of control for generation, energy storage and devices in the home. When the power goes out, Span automatically islands the house and allows the homeowner, via a cell phone, to prioritize energy use for appliances and connected devices, so that the most important are energized first.

This is different from today's traditional electrical panels, which provide only passive current protection, the smart panel can monitor and control 32 circuits. The panel also simplifies installation of solar, energy storage and electric vehicle charging — common elements of home nanogrids.

Radiant heating and radiant COOLING via panels in the ceiling.

— https://www.youtube.com/watch?v=rPC0_LxlwIY
  Infrared wallpaper.

— https://octopus.energy/blog/nexgen-electric-wallpaper/

Your House Under a Dome

Take wind out of the equation for home heating and add in all-year round solar warming. It is like having a storm door on your whole house

– https://www.theweathernetwork.com/ca/news/article/
   dome-russian-family-try-out-experimental-living-situation-yakutia
When temperatures in Yakutsk plummeted to –50 °C (–58 °F) the temperature under the dome was –34 °C. A 16C shift is a huge benefit.

• Generation

Wind Turbine

Alpha 311 vertical axis wind micro- generators pick up wind created by passing vehicles and generate up to 6 kWh a day. They plug into existing wiring on lamp posts or household wiring and mount basically anywhere (Fig. 1).

Hybrid Solar PV and Hot Water

The author is convinced there is no greater opportunity for harvesting energy for homes and small micro-grids than some type of system which creates both electrical energy and hot water from the same panel. If something like the two products (one British, one French) below can be made practical, it will lower

Fig. 1   Pole mounted micro-turbines
https://alpha-311.com/#home

the price/performance threshold necessary to make widespread rooftop systems justifiable.

These would complement geothermal storage extremely well.

Not only do the systems produce hot water but they also remove heat from the solar cells which increases their efficiency. Yes, they are more complex.

- Naked Energy (Fig. 2)

Business and industry require high-grade heat for sanitary hot water, heating and industrial processes. As **VirtuPVT** uniquely benefits from vacuum tube technology, for improved thermal performance, it can deliver a **peak efficiency of 80%**, converting 20% of the sun's energy to electricity and 60% as heat.

– https://www.youtube.com/watch?v=vzRTtaDVVW0

- Dual Sun

Hybrid efficiency up to 85% for a 25% increase in cost produce 3 units of heat output for every unit of electricity output.

– https://dualsun.com/en/
   - https://dualsun.com/en/product/hybrid-panel-spring/

Fig. 2 Hybrid solar hot water and PV systems in narrow segments

Cooling the photovoltaic cells improves electricity output by 5–15% depending on usage.

As it circulates through the exchanger, the water warms up by recovering the heat emitted by the photovoltaic cells. This water can reach a temperature of up to 70 °C and can therefore be used to cover the building's various heating needs. (or stored underground) (Fig. 3).

Do they work – almost certainly.

Are they heavy – yes, but how much can your roof bear?

Are they reliable? How well do they stand up to extreme climate?

- Electricity from Thin Air

Exploiting the voltage differentials of altitude maybe possible but is it scalable and what is the EROI?

– https://www.euronews.com/living/2020/02/18/ scientists-have-worked-out-how-to-generate-electricity-from-thin-air

Micro-grids and district heating are the future for small communities and even a handful of buildings. The Low Carbon Hub project in the UK has been established to provide the tools to enable a community to organize itself and put together systems to meet their unique needs.

– https://www.lowcarbonhub.org/

- Storage

  – Chemical Battery Champion

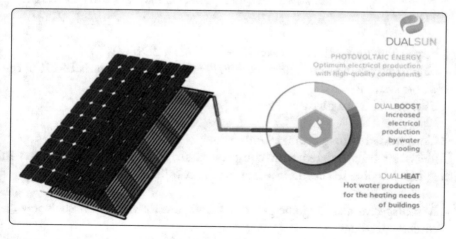

Fig. 3 Hybrid solar hot water and PV in larger panels

This isn't new technology but it's worth mentioning. This battery has been running for 176 years.

– https://www.youtube.com/watch?v=UtQGYz4f3YQ

• Deep Drilling Geothermal
The deepest hole we have ever dug illustrates the problems with very deep drilling.

– https://www.bbc.com/future/article/20190503-the-deepest-hole-we-have-ever-dug

• Ammonia
Ammonia is viewed by some as an alternative to hydrogen. It has some advantages and some disadvantages compared to hydrogen but, in the end, it is an energy carrier process which is fairly lossey.
The ammonia economy - https://www.youtube.com/watch?v=5Y_2Z_VwFNc

– Robert Service article https://www.sciencemag.org/news/2018/...
– Douglas MacFarlane Paper https://www.cell.com/joule/pdf/S2542-...

• Gravity Storage
It doesn't have to be practical for general application but might be suitable for the narrow and specific purpose of shaving peaks off generation spikes and filling in dips. These might last for only several minutes until other, slower reacting storage mechanisms with much greater capacity could be brought on line (Fig. 4).

– https://www.youtube.com/watch?v=lh1%2D%2DftWWvY
– Gravitricity.com – efficiency – 80–90% efficiency but what is EROI of the whole system?
– Energyvault.com picture

• Compressed Airbag Storage Underwater
Underwater bags inflated with compressed air. What are the heat losses in this? Is it possible to capture the heat of the compression process?

– https://spectrum.ieee.org/energy/renewables/hydrostor-wants-to-stash-energy-in-underwater-bags

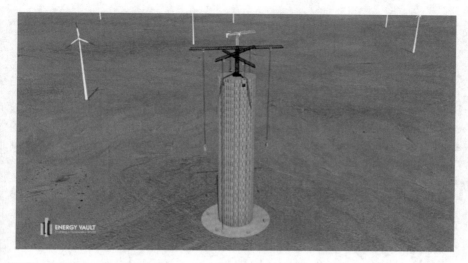

Fig. 4 Gravity storage using a cranes and concrete weights can be installed almost anywhere

- Liquid Metal Batteries
  A 500 °C working temperature yet 80% efficiency claim? It is tough to contain heat of that level without high losses.

  – https://www.youtube.com/watch?v=VNCC8QGy_u0

- Iron/air batteries for Grids – in effect, reversible rusting
  Not suitable for small mobile batteries but designed for large stationary grid storage. It is said to be capable of storing electrical energy for 4 days with 1/3 cost of lithium.

  – https://www.youtube.com/watch?v=UDjgSSO98VI
  – https://formenergy.com/technology/

- Wind
  Sailing ships; remember those? Wind powered ocean liners and super tankers. Large, conventional transport ships can use an average of 40 tons of fuel per day, generating 120 tons of CO_2 -- equivalent to driving a car 270,000 miles.
  With a projected top speed of about 10 knots, Oceanbird will be slower than standard car carriers, which can travel at 17 knots. It will take around 12 days, instead of the standard seven, to cross the Atlantic.

– https://www.cnn.com/travel/article/oceanbird-wind-powered-car-carrier-spc-intl/index.html

- Tidal Generator
This Orbital Marine generator has a similar capacity factor to wind turbines but with a shorter lifespan. What are the maintenance issues? Are its costs higher? Can it fill in the output dips of other variable sources predictably? If so, reliable output at predictable times is a rare and highly valued trait of any renewable energy source particularly at northern latitudes.

– 680 tons of machinery with a 15 year lifespan, 2 MW capacity and a capacity factor of 30% (Fig. 5).

- Nuclear / Fusion
Big nuclear and small nuclear both have similar issues which are the cost and the legacy and safety issues. And oh yes, what is the public acceptance of having one in their back yard?
SMR Small Modular Reactors Benefits:

– Less on-site construction,
– Increased containment efficiency,

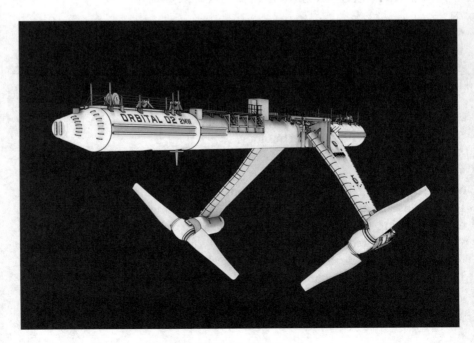

Fig. 5 Tidal power for Scotland from Orbital Marine

- Enhanced safety
- Dimensions 25 m high by 5 m wide
- Output capacity – 60 MW, capacity factor – 92%,
- Cooling pool 50 million liters of water for 12 units with 720 MW capacity
- 1 Candu reactor has capacity of 880 MW
- Cost of a 12 reactor system $4 billion CDN

- Thorium Reactors

  - Costs about the same on a kWh of capacity basis as wind but runs all of the time
  - Produces 2 kWh of heat for every 1 kWh of electricity
  - Fusion? 60 years of development and so far not kWh One of net output.
  - https://www.youtube.com/watch?v=mNlggLdWUng

- Hydrogen

Plasma Kinetics uses a solid state hydrogen process to capture gas from smokestacks. It is claimed to be 30% lighter, 7% smaller and 17% less expensive than a lithium-ion battery per kWh and doesn't use compression stages which drag down the efficiency of conventional hydrogen.

- https://plasmakinetics.com/

No matter how brightly the new technology shines, you must be able to look past the glitter and define the full lifecycle. Keep the example of the oil sands and conventional oil in mind.

The legacy costs of three million old oil wells in the USA, two million of which are unplugged is staggering.

- https://abcnews.go.com/US/wireStory/forgotten-oil-gas-wells-linger-leaking-toxic-chemicals-79,188,255?cid=clicksource_4380645_7_heads_posts_headlines_hed

Strategic Nation Re-building

Technology of Localization and Re-industrialization

Vital medical supplies and renewable energy components need to be made locally to assure reliable supplies. Some countries are trying to re-build their industrial base one industry at a time.

Is France's bike industry on the road to a 'Made in Europe' comeback? Euronews reports that "Decades of industrial relocation to countries with low production costs have caused damage to many regions in France. In the small town of Revin, closed shops, houses for sale and industrial wastelands are just some of the signs of this decline."

This will sound familiar to anyone living in the western world whose career prospects, personal finances and community were devastated by globalism. It appears some politicians in France are actually supporting local industry including the head of the French municipalities association. Bernard Dekens stated "yes, "it's obvious that we cannot go on like this, selling our industry to China and South East Asia"; "we must protect ourselves."

https://www.euronews.com/2021/07/16/is-the-french-bicycle-industry-on-the-road-to-a-made-in-europe-comeback

Not all great tech is high tech. Learning how to live simply and well is the greatest technology of all because any technology can be brought down by natural events over which we have no control.
Simplicity is a vital technology.

- High Tech Disasters

The Carrington Level Events - solar storms.

"If the 1921 storm occurred today, there would be widespread interference to multiple technological systems, and it would be quite significant," with effects including blackouts, telecommunications failure, and even the loss of some satellites, Love says. "I'm not going to say it would be the end of the world, but I can say with high confidence that there would be widespread disruption."

- https://www.scientificamerican.com/article/new-studies-warn-of-cataclysmic-solar-superstorms/
- https://www.youtube.com/watch?v=sBxjwzKwVl0

Great Sources

Videos

- Can you really affect climate change? Yes. You can. Here's how... Just Have a Think

  - https://www.youtube.com/watch?v=fRTic6mjCFk

- A Reality check on renewables – David MacKay

  - https://www.ted.com/talks/david_mackay_a_reality_check_on_renewables?language=en
- Too Clever by Half, But not Nearly Smart Enough – William Rees

  - https://www.youtube.com/watch?v=YnEXEIp5vB8&t=4352s
- Fully Charged Show on youtube – basically all things electric and many energy conservation tips.

Books

- Energy and the Wealth of Nations - Charles Hall
- The Energy of Slaves – Andrew Nikiforuk
- "Sustainable Energy – without the hot air" at http://withouthotair.com/

  - A great reference and it is free.
- Global Crisis – Geoffrey Parker
- Failing States, Collapsing Systems: BioPhysical Triggers of Political Violence - Nafeez Mosaddeq Ahmed

Games

- Game on – you and your children can build your own sustainable society – an app from https://climategame.eu/

Climate Data

- Dr. James Hansen has created a tremendous source of accessible climate data.
- http://www.columbia.edu/~mhs119/
- Insurance Study of Climate Related Losses
- https://www.insuranceinstitute.ca/en/resources/insights-research/Climate-risks-report

Modelling

- WhatIf Technologies http://www.whatiftechnologies.com/

Index

© The Author(s), under exclusive license to Springer Nature Switzerland AG 2022
J. E. Meyer, *The Post-Pandemic World*, https://doi.org/10.1007/978-3-030-91782-1